ORGANIC CHEMISTRY I LECTURE TEMPLATES

Fourth Edition

Kay I. Kouadio, Ph.D

North Lake College

Name_____

Kendall Hunt
publishing company

Cover image © Shuttertock.com

www.kendallhunt.com
Send all inquiries to:
4050 Westmark Drive
Dubuque, IA 52004-1840

Copyright © 2009, 2011, 2016, 2019 by Kendall Hunt Publishing Company

ISBN 978-1-5249-9686-4

Published in the United States of America

To all my wonderful students (past and current) who have taught me so much how to teach them.

OCHEM I: TEMPLATES - TEXTBOOK RELATIONSHIP

Textbook Chapter #	Template Unit #	Test #
	1	1
	2	1
	3	1
	4	2
	5	2
	6	2
	7	3
	8	3
	9	3
	10	4
	11	4
	12	4
	13	5
	14	5
	15	5
	16	5

TABLE OF CONTENTS

1. OCHEM UNIT 1: INTRODUCTION TO ORGANIC CHEMISTRY
A. WHAT IS ORGANIC CHEMISTRY?

- **Organic Chemistry** (Ochem) is the chemistry of carbon compounds. Other elements, such as H, O, N, S, P, Si, and the halogens, are also encountered in organic compounds.

B. THE BIRTH OF ORGANIC CHEMISTRY: A TIMELINE

- 1770: Swedish Bergman proposed that:
  -Inorganic compounds are from minerals.
Ex. NaCl
- Organic substances are from living things.
Ex. Urea

C. THE VITALISTIC THEORY

- Organic substances cannot be prepared in the laboratory.
- A **biological organ** is needed to make organic substances.

D. FRIEDRICH WÖHLER AND THE DEMISE OF THE VITALISTIC THEORY

- 1828: a fatal blow to the vitalistic theory.
- German scientist Friedrich Wöhler synthesized urea (an organic substance) from ammonium cyanate (an inorganic substance) in the laboratory without the "help of a kidney"as follows:

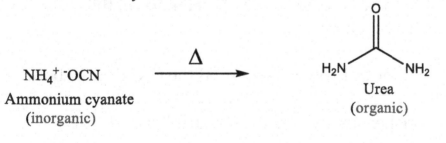

$NH_4^+\ {}^-OCN$
Ammonium cyanate
(inorganic)

Urea
(organic)

- Thereafter, the vitalistic theory was abandoned.

E. A UNIVERSE OF ORGANIC SUBSTANCES

- Nowadays, more than 10 million organic substances have been discovered or synthesized. For instance, all molecules essential for life are organic in nature: for example, DNA, lipids, carbohydrates, proteins.

- Most drugs (whether legal or illegal) are also organic substances: tylenol, aspirin, ibuprofen, AZT, cocaine, THC (the active ingredient in marijuana), heroin.

- See Fig. _____, page _____ for examples.

- Read the textbook prologue.

- **Conclusion:** OChem is **very important** in one's education.

F. DEVELOPMENT OF CHEMICAL BONDING THEORY

- In the 1800s, OChem was in its infancy. Several things we take for granted today were not well understood or had not yet been discovered. But today, because of the hard work of many eminent scientists, our understanding of the chemical bond is far better than that of our predecessors.

- The tetravalent nature of C was discovered by August Kékulé and Archibald Couper. Indeed, **carbon always forms four bonds in its compounds**. August Kékulé is also credited with the discovery of **catenation**, the ability of C to form **carbon chains**.

- Alexander C. Brown discovered the C-C double bond.

- Emil Fisher discovered the C-C triple bond.

- Finally, the tetrahedral structure of four-bonded carbon was discovered by Jacobus van't Hoff.

- **Note: Always consult** *List of How To's, List of Mechanisms on pages* **xxiii and xxiv when you are not sure of what you are doing.**

OCHEM UNIT 2: STRUCTURE AND BONDING

A. ATOMIC STRUCTURE: A REVIEW

1. **MODERN VIEW ON THE MICROSCOPIC STRUCTURE OF MATTER**
   - **Atomic Structure: A Summary**

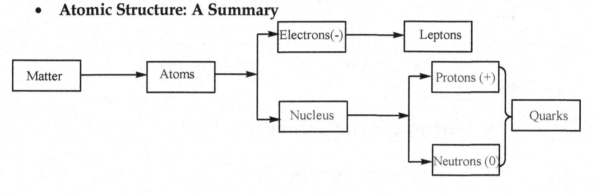

   - See Fig. on page _____.

2. **SOME DEFINITIONS**

   - **Atomic number (Z)** = number of protons in the nucleus of an atom.

Ex.

   - **Mass number (A)** = protons + neutrons.

Ex.

   - **Symbol of an atom.**

Ex.

   - **Isotopes** = atoms of the same element with different mass numbers.

Ex.

   - **Atomic weight of X** = average weight or mass of all isotopes of a given element X in amu.

Ex.

5

- **Note**: in the **neutral atom**, number of electrons = number of protons.

- See Fig. on page _____.

- **Cation** = positive ion.

Ex.

- **Anion**= negative ion.

Ex.

B. THE PERIODIC TABLE

1. INTRODUCTION

- Discovered by Mendeleev and Lothar Meyer in **1869**.
- Column = group or family. Elements in a group have similar chemical and physical properties.
- Row = period.

2. IMPORTANT ELEMENTS IN OCHEM

- C, H, O, N, Si, S, P, Li, Na, Mg, K, and halogens are the most encountered elements in OChem.

- See Fig. _____ page _____.

C. ATOMIC STRUCTURE: THE SHELL MODEL AND ORBITALS

1. INTRODUCTION

- From **wave mechanics**, electrons in atoms are in energy levels called **shells**. In a summary:

Shells ⟶ subshells ⟶ orbitals (electron densities)

2. TYPES OF SUBSHELLS
a. s subshells
- spherical
- one s orbital

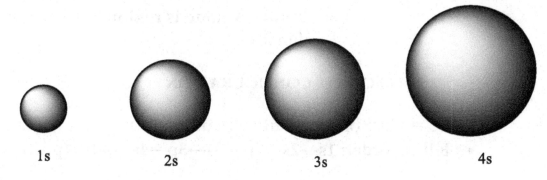

1s   2s   3s   4s

- can take up to 2 electrons

b. p subshells
- spherical lobes or dumbbell shape
- set of 3 **degenerate** orbitals (**Px, Py, Pz**) in a given subshell

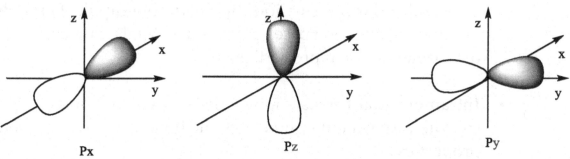

Px   Pz   Py

- can take up to 6 electrons

c. d subshells
- various shapes
- set of 5 d orbitals in a given subshell
- can take up to 10 electrons

d. f subshells
- various shapes
- set of 7 f orbitals in a given subshell
- can take up to 14 electrons

- See Fig. on page _____.

- Note: Only s and p orbitals are involved in OChem
- Recall: A node is region with zero electron density.

3. ELECTRON CONFIGURATION

- EC = Electron distribution in atom.
- Filling order: $1s \rightarrow 2s \rightarrow 2p \rightarrow 3s \rightarrow 3p \rightarrow 4s \rightarrow 3d \rightarrow 4p$............

Ex. C, H, O, N, F, Na, Cl

D. THE NATURE OF THE CHEMICAL BOND

1. INTRA AND INTERMOLECULAR FORCES: A REVIEW

- **Intermolecular forces** = ion-dipole, dipole-dipole, H bonds, London Dispersion Forces (or van der Waals' Forces): determine physical properties

- **Intramolecular forces** = ionic bonds, covalent bonds, coordinate covalent bonds, metallic bonds: determine chemical properties

- Note: Covalent bonds are the bonds mostly encountered in organic compounds.
- Recall: **metal + nonmetal = ionic compound**
  nonmetal + nonmetal = covalent or molecular compound.

2. VALENCE SHELL ELECTRONS

- **Outermost shell** electrons

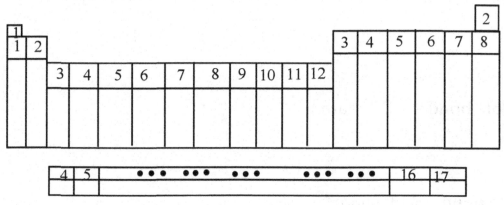

Ex: Give for each element the number of valence electrons

Element	C	H	O	F	N	S	Br	Be	B
Valence electrons									

- **Do Problem page** _____.

- **Note: Only valence electrons are involved in chemical reactions.**

3. THE COVALENT BOND

- Results from the **sharing of a pair** of valence electrons by **2 nonmetal atoms.**

- **Why?** Each atom is seeking an **octet**. Each atom wants to adopt a noble gas electron configuration = **The Octet Rule.**

Ex: Cl_2

4. TYPES OF COVALENT BONDS

- **Single bond** = only **one** electron pair is shared

Ex:

- **double bond** = 2 EP shared

Ex:

- **triple bond** = 3 EP shared

Ex:

5. GENERAL WAY OF PREDICTING THE NUMBER OF BONDS AN ELEMENT FROM THE MAIN-GROUP ELEMENTS CAN FORM

a. Atoms with n valence electrons (n ≤ 3)

- Form **n** bonds.

Ex: Give the number of bonds for each element

Element	Al	H	Mg	Li	Na	B	Be	Ba	K
# val. el.									
# bonds									

b. Atoms with n valence electrons (n ≥ 4)

- Form 8-n bonds.

Bonds = 8-n

Ex: Give the number of bonds for each element

Element	C	N	O	F	Cl	Br	S	P	I
# val. El.									
# bonds									

6. SUMMARY ON THE NUMBER OF POSSIBLE BONDS FOR SOME ELEMENTS ENCOUNTERED IN OCHEM

Element	usual # of bonds	possible
H,F	1 (always)	0
C	4	3
O	2	1,3
S	2	4,6
Cl, Br, I	1	3,5
B	3	4
N	3	1,2,4
P	3	4,5
Si	4	....

- **Note: Become familiar with the connectivities above.**

- See Fig. _____ page _____.

- Do Problems on pages _____ - _____.

E. LEWIS STRUCTURES

1. DEFINITION

- A Lewis structure is a **representation of a molecule or molecular ion** that displays the bonding pattern in that molecule or ion.

2. GENERAL GUIDELINES ON DRAWING LEWIS STRUCTURES

a. The Central Atom

- The central atom is the atom to which the "peripheral" atoms are connected. The central atom is the atom in the center, literally.

Ex: CH_4

- **Some Hints on Central Atoms:**

- H and F are never central atoms.
- C is always a central atom
- In compounds or ions containing H, O, and a 3rd atom, the 3rd atom is usually the central atom..

Ex: H_2CO_3

b. The Octet Rule Revisited

- In drawing Lewis structures, second row nonmetal elements **C, N, O, and F** must always have an octet. Br, Cl, I always have an octet when they are "**peripheral**" atoms.

Ex: PCl_3

c. Number of Bonds or Bonding Pairs in a Molecular Compound or Ion

$$\boxed{\#\text{Bonds} = \tfrac{1}{2}(\text{max \# of possible } e^- - \text{total valence } e^-)}$$

Ex: C_2H_6

c. Steps for drawing Lewis structures: See textbook, page _____.

    i.   Sum up all valence electrons in the molecule or ion.
    ii.   Identify the central atom.
    iii.   Calculate the number of bonds or bonding pairs.
    iv.   Connect all peripheral atoms to the central atom with single bonds first, making sure to respect the **connectivities** of all atoms involved.
    v.   Complete the octet for each peripheral atom.
    vi.   Place left-over electrons as pairs on the central atom.
    vii.   If there are still some electrons left, try multiple bonds for **C, N, O, S, or P**.

- **Note: For ions, a (+) or (-) charge shows the number of electrons lost or gained by the "system" during ion formation.**

Ex: Draw Lewis structures of: CCl_4, H_2O, CO_2, NH_3, NH_4^+, CO_3^{2-}

- **Read pages _____ - _____.**

- **Do Problems ____ pages _____._____**

e. Lewis Structures of Organic Substances

Ex: C_3H_8, C_2H_4, C_4H_{10}

3. FORMAL CHARGE

- The formal charge (FC) of an atom X in a molecule or ion is the charge that X would have if all atoms in the molecule have the same **electronegativity**. The FC is the charge of an atom in a Lewis structure.

$$\boxed{FC_X = (\text{\# val } e^- \text{ in X}) - \tfrac{1}{2}(BE) - (NBE)}$$

Ex: Cl_2, PH_4^+

- Do Problems on pages _____.

4. FORMAL CHARGES OF O, C, AND N IN COMPOUNDS AND IONS

element	# bonds	formal charge
O	2	0
O	1	-1
O	3	+1
N	3	0
N	4	+1
N	2	-1
N	1	-2
C	4	0
C	3	-1 or +1

- See Table _____, page _____.

5. ISOMERS

- In general, **isomers are 2 or more acceptable arrangements** of atoms in a Lewis structure.

Ex:

$C_2H_2Cl_2$

C_2H_6O

- Note: Isomers have same molecular formula, but different arrangements of atoms or different connectivities. They can have constitutional isomers, stereoisomers, etc., as we'll see in Unit 5.

- Do Problem _____, page_____.

6. EXCEPTIONS TO THE OCTET RULE: 3

a. Odd Number of Valence Electrons

Ex:

b. The Incomplete Octet: Be, B, and Al Compounds

Ex:

c. The Expanded Octet: Observed with elements of the p block from the 3rd row and beyond.

Ex:

- Read page _____.

F. RESONANCE STRUCTURES

1. DEFINITION

- Resonance structures are **alternative equivalent Lewis structures** that do not exist by themselves.

Ex: O_3, NO_3^-, CO_3^{2-}

- Do Problems on pages_____ _____.

2. RESONANCE HYBRID

- **A resonance hybrid is an average** of all resonance structures.
- A Resonance hybrid is more stable than any of the resonance structures.

Ex: O_3

3. RESONANCE CONTRIBUTION

a. Equal Contributors = Identical Resonance Structures

Ex: O_3

c. Major Contributor, Minor Contributor = better RS
- Minor contributor = least stable RS
- Major Contributor = better, more stable RS
- **Has more bonds and fewer charges**
- **Has smaller formal charges (positive or negative)**
- **Has more negative formal charges on most electronegative atoms**
- **All second row elements (C, N, O, F) have an octet**

Ex: O_3, CO_2

- Read pages _____ – _____.

- Do Problems _____ - _____pages _____.

G. MOLECULAR GEOMETRY

1. FLAW WITH LEWIS STRUCTURES

- Lewis structures appear flat
- Molecules have shapes that depend on bond lengths and bond angles.

Ex: CH_4

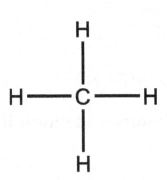

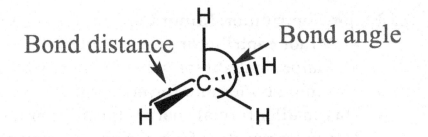

- **Read about bond lengths and trends on page _____.**

- Within a period, bond length **(H-X)** decreases from L to R.
- Within a group, bond length **(H-X)** increases from top to bottom.

- See Table _____, page _____.

2. THE VSEPR

- The VSEPR stipulates that the geometry of a molecule is determined by the number of repelling **groups** or **domains** (a **single bond, a double bond, a triple bond, or a lone pair**) **around the central atom.**
- **Groups tend to stay as far away as possible.**

Ex: CCl_4, BF_3, H_2CO, CO_2, C_2H_2

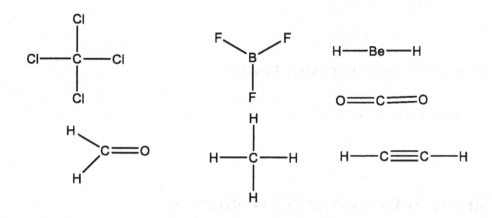

3. MOLECULAR GEOMETRY IN OCHEM BASED ON GROUPS

# of groups	geometry	bond angle
2	linear	180
3	trigonal planar	120
4	tetrahedral	109.5

Ex: Count the number of groups around each central atom in the following molecules and assign molecular geometries

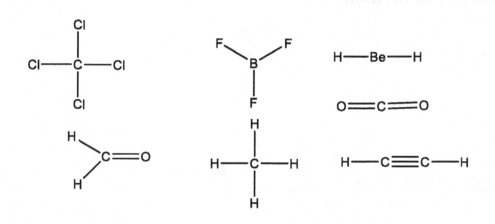

19

- Read pages _____ - _____.

- Do Problems on pages _____ - _____.

H. WRITING STRUCTURES OF ORGANIC SUBSTANCES

1. INTRODUCTION

- Recall: C = 4 bonds
- H = 1 bond

2. WAYS OF REPRESENTING MOLECULES: 4

a. Molecular Formula

Ex: C_4H_{10}

b. Structural Formula or Lewis Structure

c. Condensed Formula

d. Skeletal Formula

- Do all problems from page _____ to page _____.

I. THE VALENCE BOND THEORY AND ORBITAL HYBRIDIZATION

1. ROLE OF VALENCE SHELLS IN BONDING

- **2 theories**: Valence Bond (VB) and Molecular Orbital (MO) Theories

2. THE VALENCE BOND THEORY

- Bonding results from the **overlap of 2 singly occupied valence atomic orbitals.**

Ex: H_2

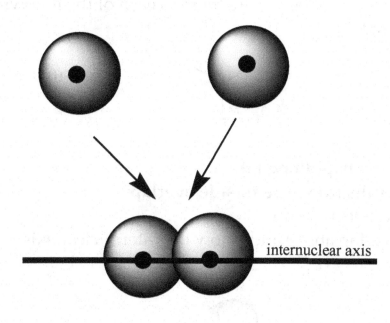

internuclear axis

Ex: Cl_2

3. TYPES OF COVALENT BONDS RESULTING FROM VALENCE ORBITAL OVERLAP

a. Sigma Bond (σ)
- results from **head to head overlap**.
- is symmetrical along the internuclear axis.
- 3 ways

- **Note: a single bond is always a σ.**

Ex: How many **σ** bonds are there in each of the following compounds?

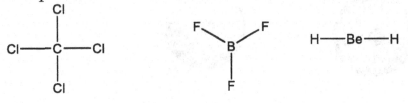

b. pi Bond (π)
- results from **side-to-side overlap**.
- parallel overlap.
- overlap above and below the internuclear axis.

- **Note: A double bond is made of 1 sigma bond and 1 pi bond.**
- **a triple bond is made of 1 sigma bond and 2 pi bonds.**

Ex: Count the number of pi bonds in each of the following compounds.

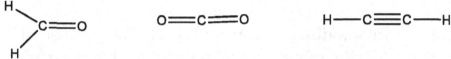

c. A Summary of Sigma (σ) and pi (π) Bonds

bond type	overlap mode	single bond	double bond	triple bond
σ	head to head	1	1	1
π	sideways	0	1	2

4. HYBRIDIZATION OF ORBITALS : sp³ and CH₄

a. Introduction

- **Question: How can 4 Hs bond to 1 C to form CH₄?**

- **Read pages _____ - _____.**

- possible because of **hybridization of C valence atomic orbitals (s and p).**
- theory introduced in 1931 by **Linus Pauling.**
- hybridization provides **stronger and more stable bonds.**

b. Steps in the hybridization process: 2 steps

      i.   Promotion of 1 e- from the s to the p subshell.

      ii.   Mixing of s and p orbitals to get 4 new sp^3 orbitals or 4 sp^3 hybrid orbitals.

      iii.   Rule: mix **n atomic** orbitals → get n **hybrid orbitals.**

c. Sp^3 hybridization

- Note: sp^3 means one s orbital on C (central atom) and three p orbitals on the <u>same</u> C atom have mixed to form that kind of hybrid orbitals. See hybrid orbitals in your textbook page _____.

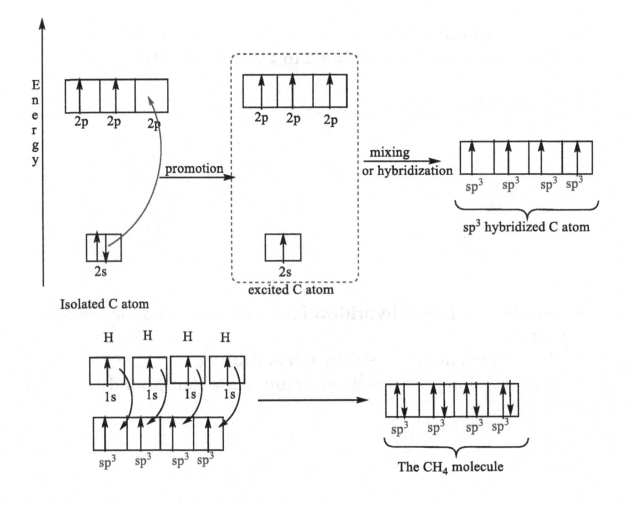

- ## The VB sketch of CH₄

d. sp² hybridization: One s orbital is mixed with **two p** orbitals on the **same central atom. One p orbital is free.**

Ex: BF₃

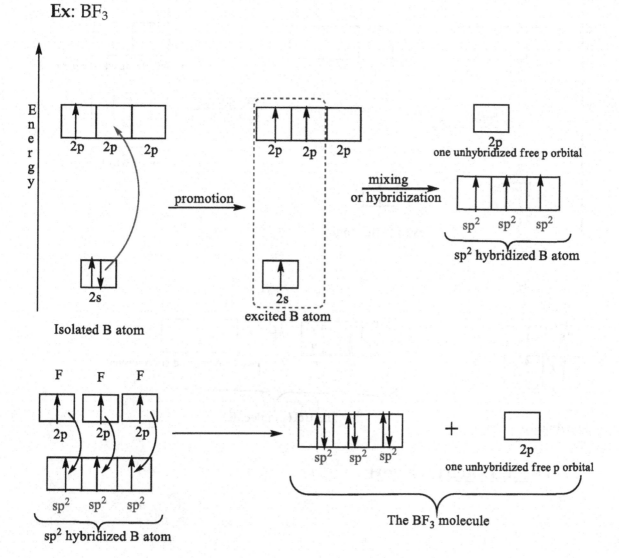

- <u>**The VB sketch of BF$_3$**</u>

e. sp hybridization: One s orbital is mixed with **one p** orbital on the **same central atom. Two p orbitals are free.**

Ex: BeH$_2$

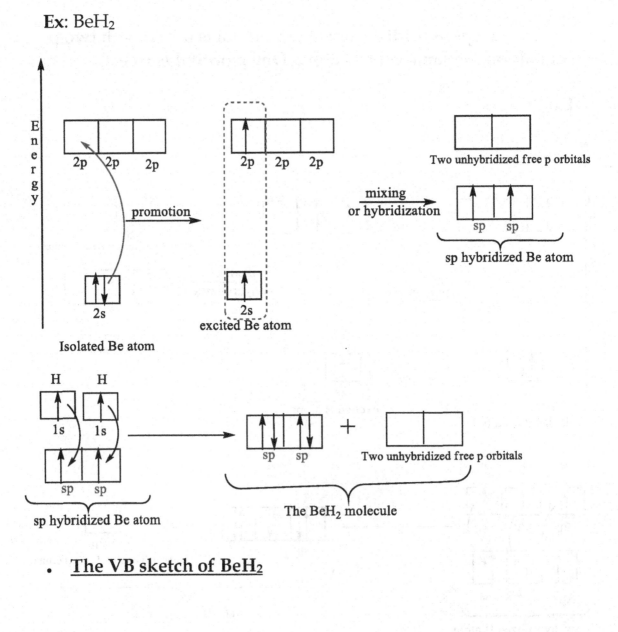

- <u>**The VB sketch of BeH$_2$**</u>

• Read pages _____ - _____

6. HYBRIDIZATION AND THE NUMBER OF GROUPS AROUND THE CENTRAL ATOM

#groups	orbitals used	hyb. orbitals	hybridization	geometry	bond angle
2	s + p	2	sp	linear	180
3	s + two p	3	sp^2	trigonal planar	120
4	s + three p	4	sp^3	tetrahedral	109.5

- See Table on page _____.

Ex: CH_4, $BeCl_2$, BF_3

7. HYBRIDIZATION OF N AND O

- See Fig. _____, page _____.

Ex: NH_3 and H_2O

8. **HYBRIDIZATION IN C$_2$H$_6$, C$_2$H$_4$, C$_2$H$_2$, AND C$_6$H$_6$:** Read pages
_____ - _____.

    a. C$_2$H$_6$

    b. C$_2$H$_4$

    c. C$_2$H$_2$

d. C_6H_6

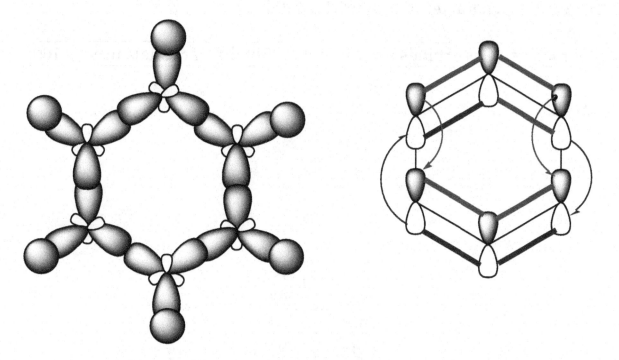

σ bonds   **π cloud (6 electrons)**

- Have 6 delocalized **π electrons (spread over 6 carbons).**

- See Table _____ for a Summary; Read pages _____ – _____ and do all related problems.

9. **HYBRIDIZATION, BOND LENGTH, AND BOND STRENGTH**

a. The C-C Bonds

bond type	C—C	C=C	C≡C
hybridization	sp³-sp³	sp²-sp²	sp-sp
BL	1.53	1.34	1.21
BE	368	635	837

- **Note: The shorter the bond, the stronger the bond, and vice versa.**
- **In a double bond, the π bond is weaker than the σ bond.**

29

b. Percent s character of a hybrid orbital

% s character = [(# s orbitals used in hybr)/(total # of orbitals used)]x100

Ex: sp^3

hybrid	%s
sp	50%
sp^2	33%
Sp^3	25%

- **Note: The higher the % s character, the stronger the bond, the shorter the bond length.**

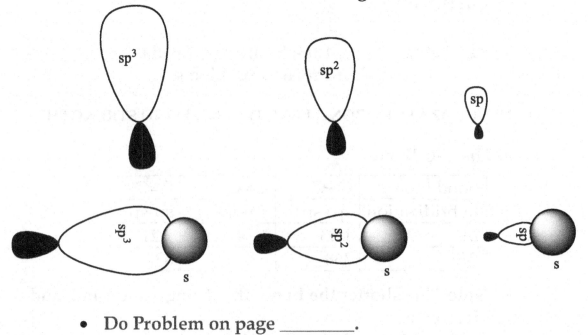

- **Do Problem on page _____.**

c. The C-H bond

Bond type	C—H	=C—H	H—C≡
hybridization	sp^3	sp^2	sp
BL	1.11	1.10	1.09
BE	410	435	523
% s character	25%	33%	50%

- **Note: Similarly to the C-C bond, the higher the % s character in the C-H bond, the stronger the bond, and vice versa.**
- **Note: Within a double bond, the π bond is weaker than the σ bond.**
- Do problem on page _____.

J. PRINCIPLES OF MOLECULAR ORBITAL THEORY (MO)

1. **BASIC IDEA**
   - **Developed by Friedrich Hund and Robert Millikan ca 1929.**
   - The **MO theory** believes that electrons in molecules and molecular ions are in the **delocalized** molecular orbitals of each individual molecule or ion, just like electrons in atoms are in atomic orbitals. Molecular orbitals result from the **combinations** (or overlaps) of the atomic orbitals in the molecule. If the atomic orbitals combine or overlap head-to-head, then a **σ MO** results. However, if the combination or overlap is "sideways", then the MO is called a **π MO**.

2. TRAVELING WAVES, STANDING WAVES

a. Traveling Waves
- A **traveling wave** is a wave that moves forward in space.

31

Ex. Light wave

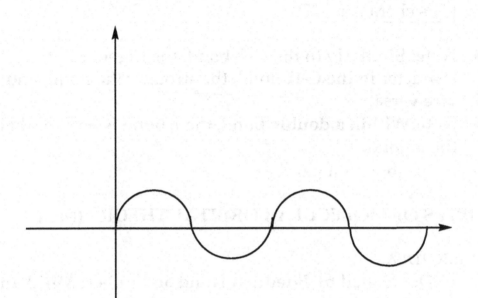

b. Standing Waves

- A **standing wave** is a fixed wave. It is a wave that is confined in a **fixed area** of space.

Ex. The vibrating string of a guitar is a standing wave.

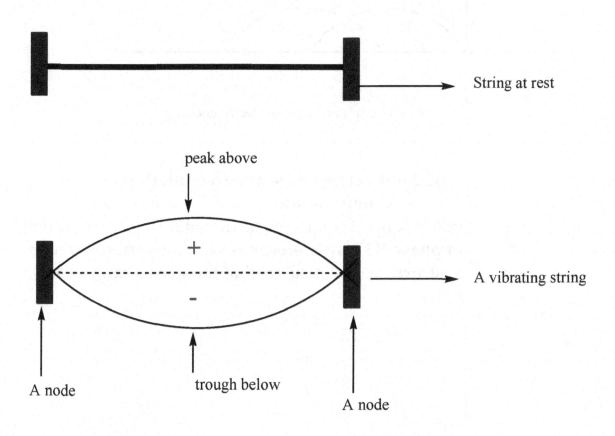

c. Interfering Waves

- In general, waves can interfere **additively** or **subtractively**.

i. Additive or Constructive Interference (Combination)
- When waves are **in phase** (they vibrate the same way), they interfere additively or **constructively**. In this case, the interfering waves **reinforce each other**. See figure below:

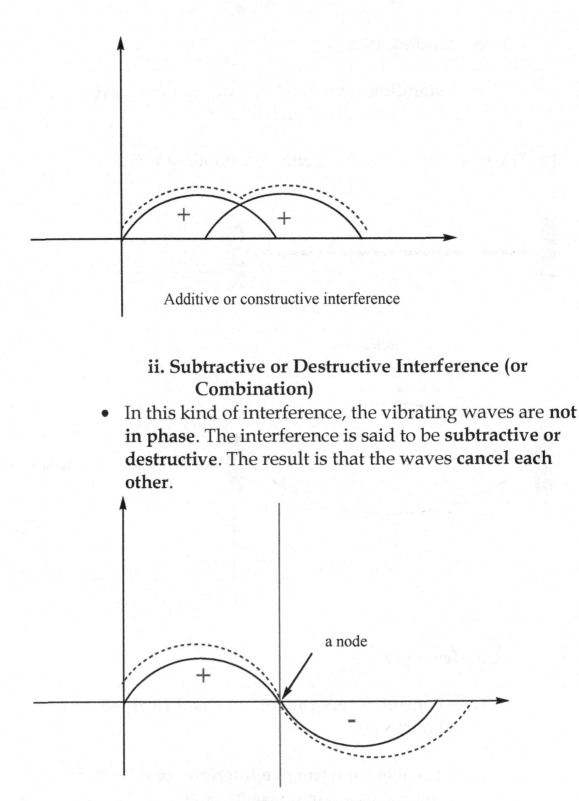

Additive or constructive interference

ii. Subtractive or Destructive Interference (or Combination)

- In this kind of interference, the vibrating waves are **not in phase**. The interference is said to be **subtractive or destructive**. The result is that the waves **cancel each other**.

a node

subtractive or destructive interference

3. APPLYING THE STANDING WAVE CONCEPT TO MOS

- **Recall: atomic orbitals (s, p, d, f) are wave functions.**
- Let's assume that these atomic orbitals are **standing waves.**

- Let's consider the **H₂ molecule:**

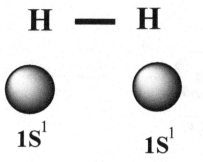

H — H

$1S^1$ $\qquad$ $1S^1$

- We can have **constructive** or **destructive** combinations (interferences) between the two s orbitals.

4. BONDING AND ANTIBONDING MOLECULAR ORBITALS (MOs)

a. MOs from two s orbitals

i. Additive (or Constructive) Combination of **two s Orbitals (in Phase)**

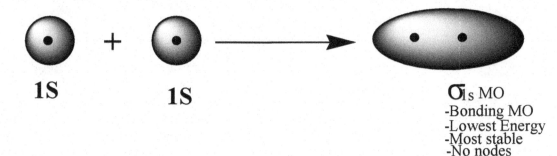

1S $\qquad$ 1S $\qquad$ σ_{1s} MO
-Bonding MO
-Lowest Energy
-Most stable
-No nodes

- → bonding mo→σ_{1s}→lowest energy→more stable→no node. **See Fig. above.**

ii. Subtractive (or Destructive) Combination of Two s Orbitals (out of Phase)

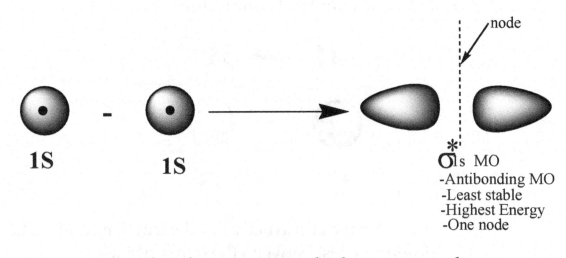

σ^*_{1s} MO
-Antibonding MO
-Least stable
-Highest Energy
-One node

- → antibonding mo→σ^*_{1s}→highest energy→least stable→1 node. **See Fig. above.**

b. Energy Level Diagram of H_2 Using MOs

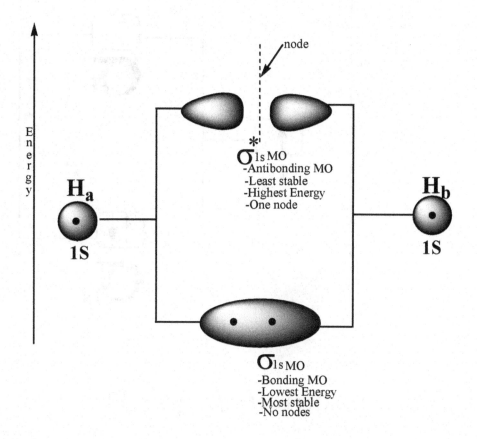

c. Electron Configuration of the H_2 Molecule

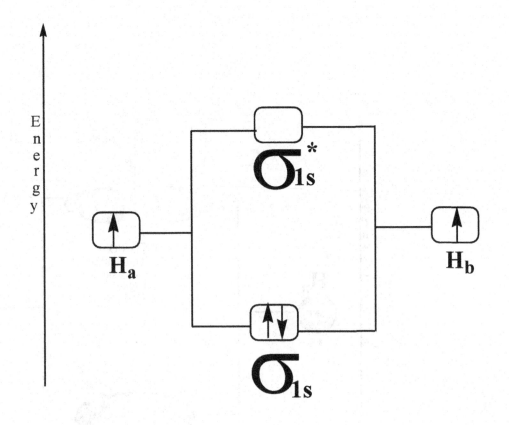

- EC of H_2: $(\sigma_{1s})^2(\sigma^*_{1s})^0$

5. THE MO THEORY APPLIED TO THE TWO π ELECTRONS IN ETHENE

a. Introduction

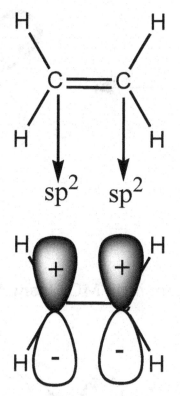

b. Π MOs from Atomic p Orbitals: "Side Ways": 2 MOs

i. Constructive or Additive Combination: In Phase

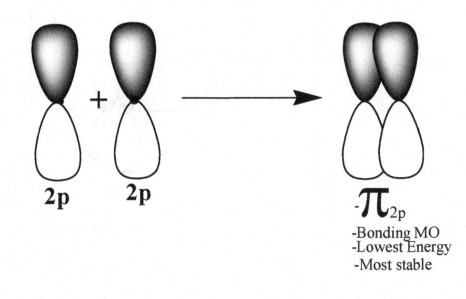

2p   2p $-\pi_{2p}$
-Bonding MO
-Lowest Energy
-Most stable

ii. Destructive or Subtractive Combination: Out of Phase

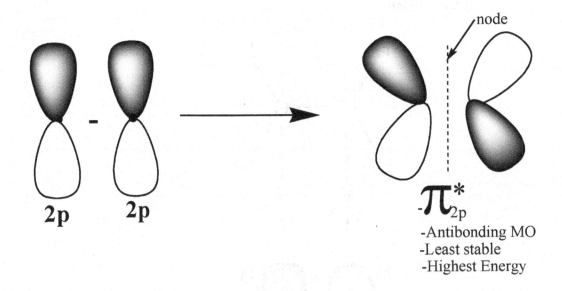

c. Energy Level Diagram of Π MOs from Atomic p Orbitals

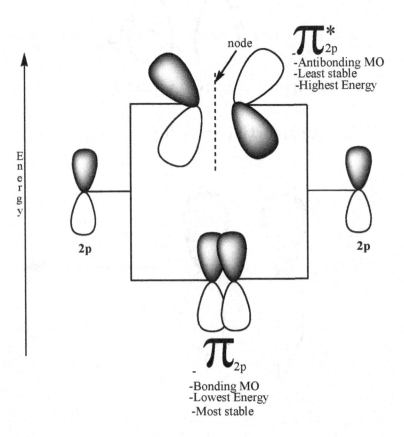

d. Energy Diagram of Ethene

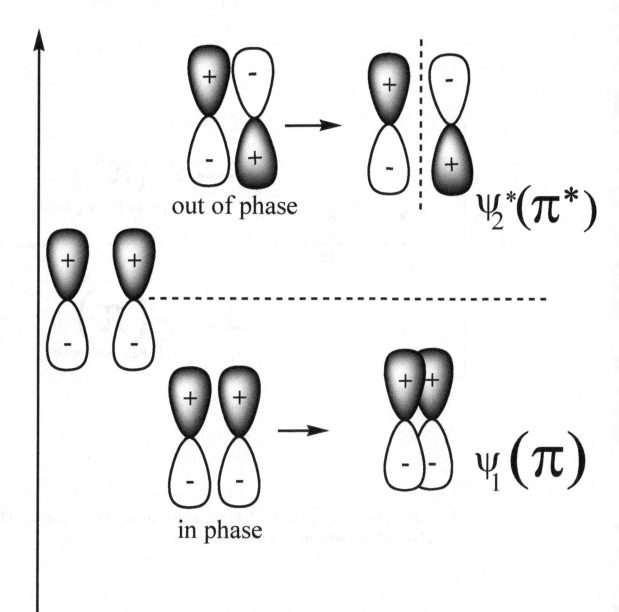

out of phase $\rightarrow$ $\psi_2^*(\pi^*)$

in phase $\rightarrow$ $\psi_1(\pi)$

e. MOs of the pi Electrons in Ethene: A Summary

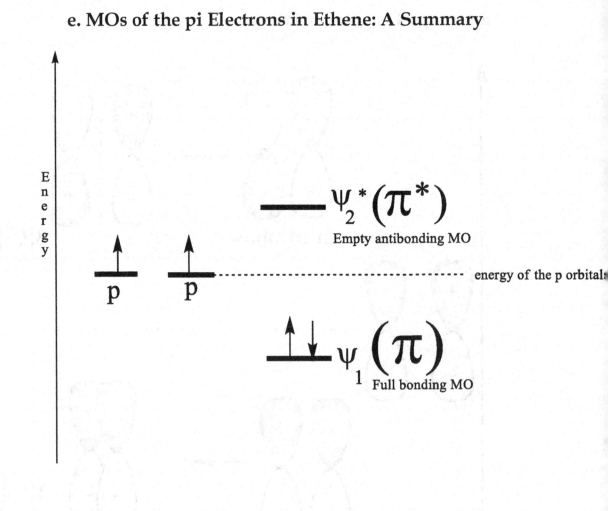

- Note: The bonding MO is full; the antibonding MO is empty. This MO will be used for hyperconjugation

6. THE MO THEORY APPLIED TO THE FOUR π ELECTRONS IN 1,3-BUTADIENE

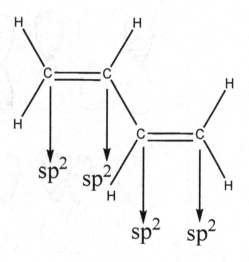

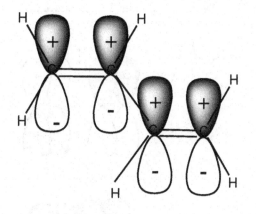

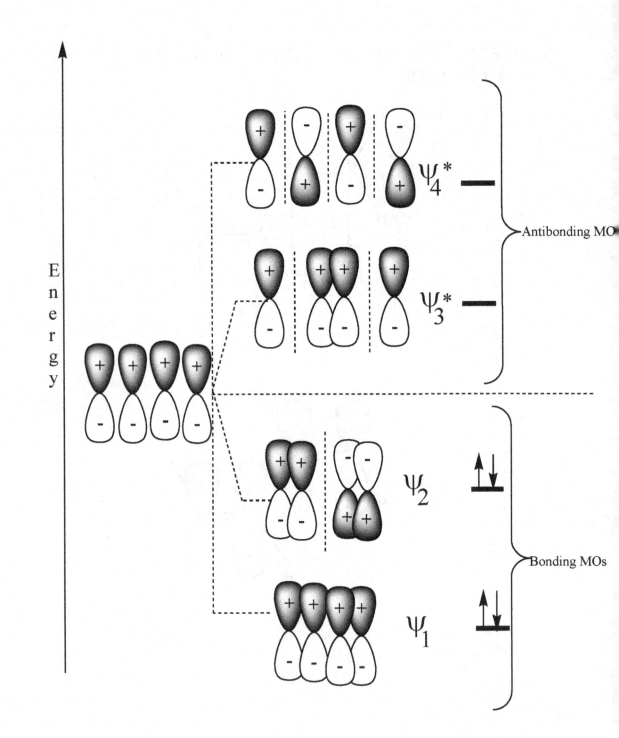

- **MOs of the 4 pi Electrons in 1,3 – Butadiene: A Summary**

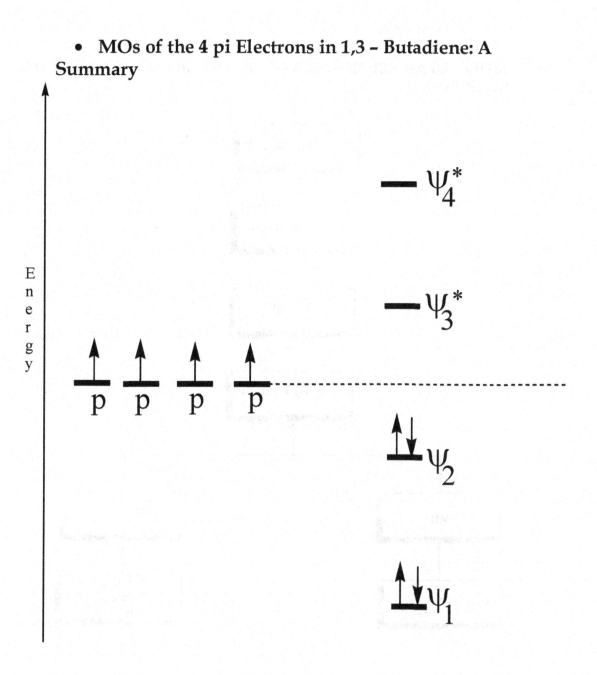

- **Note: The two bonding MOs are full; the other two antibonding MOs are empty.**

7. MOLECULAR REPRESENTATIONS AND BONDING THEORIES: A SUMMARY

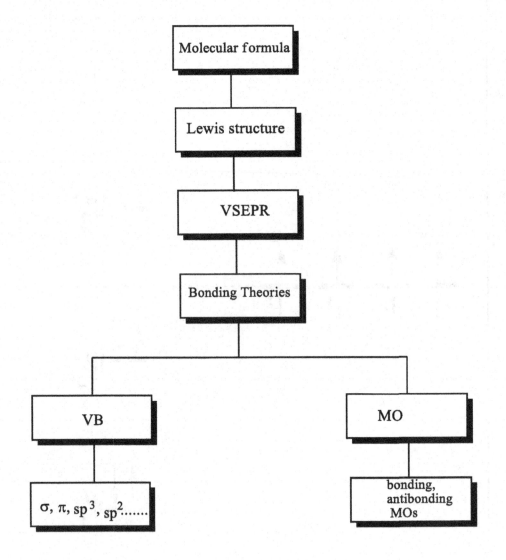

K. ELECTRONEGATIVITY AND BOND POLARITY

1. DEFINITION

- Electronegativity (EN) is a measure of the ability of an atom to attract bonding electrons to itself in a covalent bond.

Ex.

2. ELECTRONEGATIVITY SCALE: LINUS PAULING SCALE

- See Fig. _____, page _____.

Element	F	C	H	Na	Cl	Br	I	O	N
EN	4.0	2.5	2.1	.9	3.0	2.8	2.5	3.5	3.0

3. TRENDS IN EN

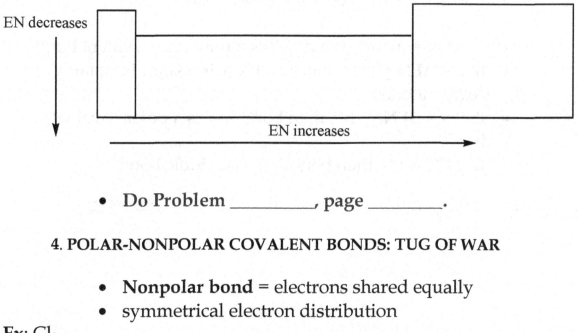

EN decreases

EN increases

- Do Problem _____, page _____.

4. POLAR-NONPOLAR COVALENT BONDS: TUG OF WAR

- **Nonpolar bond** = electrons shared equally
- symmetrical electron distribution

Ex: Cl_2

- **Polar bond** = electrons shared unequally
- unsymmetrical electron distribution

Ex: HCl

- **Note: A polar bond has a dipole moment (vector polarity).**
- **In a polar bond, there is a separation of charge.**

- **Do Problem _____, page _____.**

5. **EN DIFFERENCE AND BOND POLARITY**

- have bond X — Y

$$\Delta EN_{(X-Y)} = \mid EN_Y - EN_X \mid$$

Ex:

- **Using ΔEN to determine** bond polarity

- If $\Delta EN = 0$, then bond X-Y is a nonpolar covalent bond
- If $0 < \Delta EN \leq 0.5$, then bond X-Y is a slightly polar covalent bond
- If $0.5 < \Delta EN \leq 1.9$, then bond X-Y is a polar covalent bond
- If $\Delta EN > 1.9$, then bond X-Y is an ionic bond.

Ex: C-Cl, C-H, NaCl.

6. POLARITIES OF MOLECULES

a. Polar Molecules

- In a polar molecule, individual vector polarities **do not cancel out**. Rather, they reinforce each other.
- Molecular polarity depends on the geometry of the molecule.

Ex: H_2O, CH_2Cl_2

- See Fig. _____ and _____, page _____.

b. Nonpolar Molecule
- All bond dipoles (or vector polarities) **cancel out**.

Ex: CO_2, CCl_4, BF_3

7. ELECTROSTATIC POTENTIAL PLOTS (ESP PLOTS)

- ESP plot= map of electron density distribution in a molecule.
- obtained from calculated charge distributions.

- **Red** = electron-rich
- **Blue** = electron-poor
- **Green, orange, yellow = intermediate** electron density

- Read page _____.

- See Fig. _____, page _____.

- Do Problem _____, page _____.

- Do Problem _____, page _____.

- Review Key Concepts, pages _____ - _____.

OCHEM UNIT 3: ACIDS AND BASES

A. GENERAL DEFINITIONS OF ACIDS AND BASES

1. INTRODUCTION: 3
- Arrhenius Definition
- Brønsted-Lowry Definition
- Lewis Definition

2. ARRHENIUS DEFINITION

- An acid is a substance that can release H^+ or H_3O^+ ions in solution.

Ex:

$$HCl \longrightarrow H^+ + Cl^-$$

or

$$HCl + H_2O \longrightarrow H_3O^+ + Cl^-$$

- **An acid must contain H^+**

- A base is a substance that can release OH^- ions in solution.

Ex:

$$LiOH \longrightarrow Li^+ + OH^-$$

- A **base** must contain **OH^-**.

3. BRØNSTED-LOWRY DEFINITION

- An acid is a **proton (H^+) donor.**
- A base is **a proton acceptor.**

Ex:

$$NH_3 + H_2O \rightleftharpoons NH_4^+ + OH^-$$

- Note: An acid must contain H^+ as in the Arrhenius definition. But a base does not have to contain OH^-. A base has a lone pair of electrons or a π bond. A base may be neutral or negative.

Ex: PH_3, H_2O, NH_3, CH_3NH_2, $—C{=}O$, OH^-, $CH_2{=}CH_2$, RO^-.

- See Fig. _____, page _____.

- Note: Li^+, Na^+, K^+ are counterions = spectator ions = neutral.

4. LEWIS DEFINITION

- An acid = electron pair acceptor

Ex:

$$:F^- + BF_3 \longrightarrow BF_4^-$$

- Note: Overall, an acid must contain a proton, or have a positive charge, or have an empty valence orbital.

- A base = electron pair donor.

Ex:

$$:F^- + BF_3 \longrightarrow BF_4^-$$

- Note: Overall, a base must contain 3 "things": a hydroxide group, or a pair of electrons or a pi bond.

5. ACID-BASE DEFINITIONS; A SUMMARY

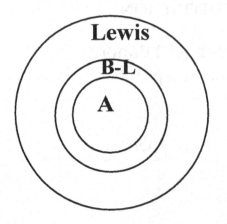

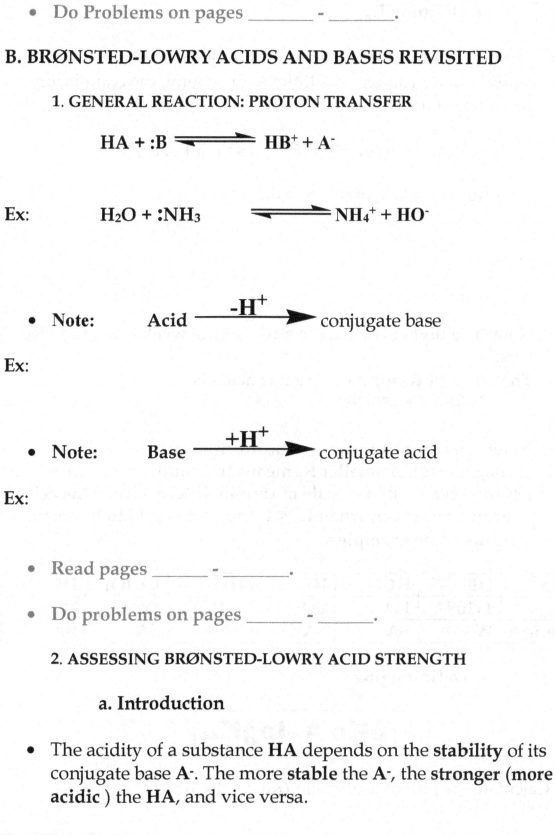

- Do Problems on pages _____ - _____.

B. BRØNSTED-LOWRY ACIDS AND BASES REVISITED

1. GENERAL REACTION: PROTON TRANSFER

$$HA + :B \rightleftharpoons HB^+ + A^-$$

Ex: $$H_2O + :NH_3 \rightleftharpoons NH_4^+ + HO^-$$

- **Note:** Acid $\xrightarrow{-H^+}$ conjugate base

Ex:

- **Note:** Base $\xrightarrow{+H^+}$ conjugate acid

Ex:

- Read pages _____ - _____.

- Do problems on pages _____ - _____.

2. ASSESSING BRØNSTED-LOWRY ACID STRENGTH

a. Introduction

- The acidity of a substance **HA** depends on the **stability** of its conjugate base **A⁻**. The more **stable** the **A⁻**, the **stronger (more acidic)** the **HA**, and vice versa.

Ex: HF and HBr

b. Defining K_a

- Most B-L acids are **weak acids**. Therefore, they reach **equilibrium** in aqueous solutions. In general, the equilibrium equation for a weak acid HA is

$$HA_{(aq)} + H_2O_{(l)} \rightleftharpoons H_3O^+_{(aq)} + A^-_{(aq)}$$

The equilibrium constant of this reaction is:

Ex: HF

- **Note: The higher the K_a, the stronger the weak acid, and vice versa.**
- **The range of K_a for most organic acids is**
        $$1\times10^{-50} \leq K_a \leq 1\times10^{-5}$$

- **Note: A larger K_a means the equilibrium lies to the right (stronger acid). A smaller K_a means the equilibrium lies to the left (weaker acid). As a rule of thumb, if $K_a \geq 1$, then the acid is strong. However, when $K_a < 1$, the acid is said to be weak.**
- **Here are some examples:**

Acid	HF	HCl	HBr	HI	$HClO_4$	$HClO_3$
K_a	$1\times10^{-3.2}$	1×10^6	1×10^9	1×10^{10}	1×10^{10}	10
Strength	WA	SA	SA	SA	SA	SA

c. Defining pK_a

$$\boxed{\textbf{pKa = -logKa}}$$

Ex: Calculate the pKa of acetic acid ($K_a = 1.8 \times 10^{-5}$).

- **Note: The lower the pK_a, the stronger the weak acid and vice versa.**

Ex: Which one is stronger: HF ($pK_a = 3.45$) or HCN ($pK_a = 9.31$)?

- Do problems on page _____.

d. Inverse Relationship

- The **stronger** the acid (**small pKa**), the **weaker** its conjugate base, and vice versa.

Ex: HF is stronger than HCN, so F⁻ is_____ than CN⁻.

- Read pages _____ - _____.

- See Table _____, page _____.

- Do problems on pages _____ - _____.

3. PREDICTION OF THE OUTCOME OF AN ACID BASE REACTION USING pKa VALUES

General Rule:

stronger acid + stronger base $\rightleftharpoons$ **weaker acid** + weaker base

Given:

$$HA + :B \rightleftharpoons HB^+ + A^-$$

Compare the pK_a of HA to the pK_a of HB⁺. pK_a (HA) **<** pK_a (BH⁺) means HA is a **stronger** acid than BH⁺, or HB⁺ is more **stable** than HA; therefore, the reaction proceeds **as written**. On the other hand, pK_a (HA) **>** pK_a (BH⁺) means HA is a weaker acid than BH⁺, or HB⁺ is

less stable than HA; therefore, the reaction **does not proceed as written.** However, the **reverse reaction** will proceed.

Ex: pK_a (CH_3COOH) = 4.76; pK_a (H_2O) = 15.74; pK_a ($H-C\equiv C-H$) = 25
Which one of the following reactions will proceed as written?

$$CH_3COOH \ + OH^- \ \rightleftharpoons \ H_2O + CH_3COO^-$$

$$H-C\equiv C-H + OH^- \ \rightleftharpoons \ H_2O + H-C\equiv C:^-$$

- **Read pages _____ – _____ and do all associated problems.**

4. GETTING Ka FROM pKa

- From the definition of pKa:

$$pK_a = -\log K_a, \ K_a = invlog(-pK_a) = 1.0 \times 10^{-pK_a}$$

Ex: The pK_a of acetic acid (CH_3COOH) is 4.74. What is its K_a?

5. FACTORS THAT AFFECT ACID STRENGTH

a. Introduction

- **Recall: The more stable the conjugate base, the more acidic the acid.**

- There are **4 factors** that determine the strength of an acid:

- **Periodic trends (or element effects)**
- **Inductive effects**
- **Resonance effects**
- **Hybridization effects**

- Read pages _____ – _____ . Do all associated problems.

b. **Periodic Trends in the Strength of Binary Acid HA (Element Effects)**

- **Have two competing factors: size and electronegativity.**

- The strength of **HA** depends on the **location** of **element A** on the Periodic Table.
- **Within a period**, acid strength **increases** with **increasing electronegativity (EN predominates over size)**. Given **H—A** ⟶ H$^+$ + **A**$^-$ and **H—B** ⟶ H$^+$ + **B**$^-$, if A is more **EN** than B, then A$^-$ is more stable than B$^-$; this means that H—A is a **stronger** acid than H—B.

Ex: H_2O and HF.

- Within a group, acid strength of **HA increases** with **increasing** size of **A (size predominates over EN)**. In this case, the negative charge of A$^-$ spreads over a larger volume (or surface area). This creates a stabilizing effect. Given **H—A** ⟶ H$^+$ + **A**$^-$ and **H—B** ⟶ H$^+$ + **B**$^-$, if A is **bigger** than B, then **A**$^-$ is more stable than **B**$^-$; this means that H—A is a **stronger** acid than H—B.

Ex: HF and HI

- Since **I**$^-$ is more stable than F$^-$, HF is a _____ acid than HI.

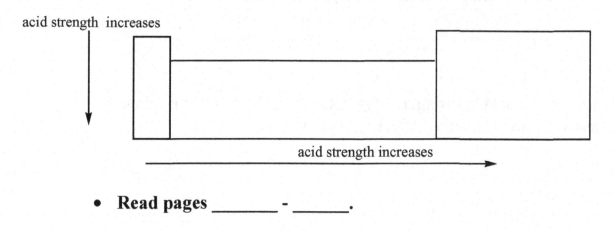

- **Read pages _____ - _____.**

c. Inductive Effects; Stabilizing Effects

- An **inductive effect** is the **withdrawal** of bonding electrons by nearby electronegative atoms (electron deficient atoms).

- Read pages _____ - _____.

- Do problems on page _____.

Ex: Explain why CH_3COOH is less acidic than CF_3COOH

- **Note: The more the electron withdrawing groups in an acid, the more acidic the acid.**

Ex:

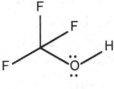

pKa = 16   pKa = 12.4

d. Resonance Effects: Stabilizing Effects

- Given **H—A** ⟶ **H⁺ + A⁻** and **H—B** ⟶ **H⁺ + B⁻**, if **A⁻** is **resonance stabilized** then **A⁻** is more stable than **B⁻**; this means that **H—A** is a stronger acid than **H—B**.

- Read pages _____ - _____, and do all problems.

Ex: Which acid is stronger, CH_3COOH or CH_3CH_2OH? Explain your answer using resonance effect.

e. Hybridization Effects

- A simple observation: $H-C\equiv C-H$ (pK_a = 25) > $H_2C=CH_2$ (pK_a = 44) > H_3C-CH_3 (pK_a = 50).
- Note: Acid strength increases with increasing %s character because the lone pair of electrons on the conjugate base is closer to the nucleus. This effect increases the stability of the conjugate base. In other words, the conjugate base with the highest % s character is the most sable. Therefore, its conjugate acid is the strongest acid.

- See summary, page _____ , and do all problems on pages _____ - _____.

6. COMMON ACIDS AND BASES USED IN OCHEM

a. Common Acids
- HCl, H_2SO_4, HNO_3 = strong acids
- Acetic acid = CH_3COOH
- p-toluenesulfonic acid = TsOH (strong organic acid))

b. Common Bases: 3 Kinds

- Oxygen bases = NaOH, LiOH, KOH, ROH (CH_3CH_2OH).
- Nitrogen bases = NH_3, NH_2^-
- Hydrides = H^- (NaH, $LiAlH_4$, $NaBH_4$)

- **Note: The conjugate base of a SA is always neutral.**

- See Fig. _____, page _____.

Ex:

$$ROH + NaH \longrightarrow RO^-\, Na^+ + H_2$$

$$CH_3CH_2OH + KH \longrightarrow CH_3CH_2O^-\, K^+ + H_2$$

- Note: ROH = an acid and NaH (KH, LiH, …..) = base.
- Read about aspirin = an acid (pages _____ - _____).

C. ACIDS AND BASES = THE LEWIS DEFINITION Revisited

1. INTRODUCTION

- **Broader definition**
- acid = electron pair acceptor
- base = electron pair donor

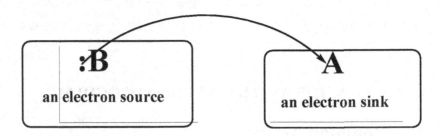

- **Note: a Lewis base is called a nucleophile (positive liking)**
- **A Lewis acid is called an electrophile (electrons are negative liking)**

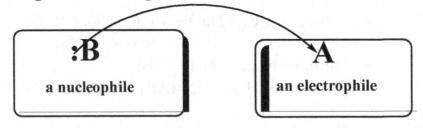

2. LEWIS ACIDS: 3 KINDS

- Neutral proton donors: inorganic acids, carboxylic acids, alcohols.

Ex: H_2O, HCl, CH_3CH_2OH

- metal cations: **empty orbitals** that can accept electrons.

Ex: Mg^{2+}, Li^+, Ba^{2+}, Fe^{3+}

- Compounds of group **3A elements and other transition metals (empty orbitals).**

Ex: $AlCl_3$, BF_3, $TiCl_4$, $FeCl_3$, $ZnCl_2$, $SnCl_4$...

$AlCl_3$ → empty p orbital

3. LEWIS BASES

- Compounds or ions containing O, N, and S atoms.

Ex: CH_3SH, HS^-

- Do problems pages _____ – _____.

- See Key Concepts, page _____.

OCHEM UNIT 4: INTRODUCTION TO ORGANIC COMPOUNDS AND FUNCTIONAL GROUPS

A. INTRODUCTION TO HYDROCARBONS

1. DEFINITION

- A hydrocarbon is an organic compound containing only H and C atoms.

Ex:

2. CLASSES OF HYDROCARBONS

- There are **3 major classes** of hydrocarbons based on the **carbon-carbon bond. See Table _____, page _____.**
- **Saturated hydrocarbons** contain only single carbon-carbon bonds. In a saturated hydrocarbon, there are no multiple bonds (no double or triple bonds).
- **Unsaturated hydrocarbons** contain at least one **multiple bond** (a triple bond or a double bond or both).
- **Aromatic hydrocarbons (or arenes)** are a special class of unsaturated ring compounds related to **benzene**.

3. SATURATED HYDROCARBONS

- Contain only **C-C single bonds**. In other words, all carbon atoms are sp^3 hybridized in a saturated hydrocarbon.

Ex:

- There are **2 types** of saturated hydrocarbons:

- **The alkanes** or **aliphatic hydrocarbons** = **acyclic** = **paraffins**.
Ex:

- **Cycloalkanes** = **cyclic**.
Ex:

4. UNSATURATED HYDROCARBONS

- contain at least one **C-C multiple bond**.
Ex:

- There are **3 types** of unsaturated hydrocarbons:
- **Alkenes = Olefins =acyclic** compounds that contain at least one C-C double bond.
Ex:

- **Cycloalkenes** = **cyclic** compounds that contain a **C-C double bond**.
Ex:

- **Alkynes** = **acyclic** compounds that contain at least a **C-C triple bond**.
Ex:

5. AROMATIC HYDROCARBONS (ARENES)

- This is a special class of unsaturated **cyclic compounds** related to the structure of **benzene**.

Ex:

6. CLASSES OF HYDROCARBONS: A SUMMARY

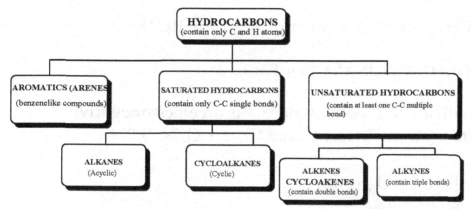

- Note: Alcohols, ethers, carboxylic acids, ketones, aldehydes, etc., are derivatives of hydrocarbons.

- Read pages _____ – _____.

B. FUNCTIONAL GROUPS

1. DEFINITION
- The functional group of an organic compound is the **face**, or **characteristic unit**, in that substance.

- See Table _____, page _____.

- See Table _____, page _____.

2. REPRESENTATION OF MOLECULES CONTAINING FUNCTIONAL GROUPS

Ex: Ethanol = CH_3CH_2OH

- $R = CH_3CH_2-$ and OH = functional group: **ROH**

3. TYPES OF FUNCTIONAL GROUPS: 3

- Entities with a **C-Z σ** bond, where Z is an **electronegative heteroatom** (atom different from C). Z = O, N, S, X (X = halogen)
- Entities with a **C=O** group.
- hydrocarbons

4. MOLECULES WITH A C-Z σ BOND

a. General Structures

C-Z

Ex:

b. Halides or Halo Functional Groups

R-X

Ex:

c. The Alcohol Functional Group (or Hydroxy Group)

R - OH

Ex:

d. The Ether Functional Group (Alkoxy Group)

R – O – R′

Ex:

e. The Amine Functional Group (Amino Group)

$R - NH_2$ or $R_2 - NH$ or $R_3 - N_3$

Ex:

f. The Thiol Functional (Mercapto or Thio Group)

R – SH

Ex:

g. The Sulfide Functional (Alkylthio Group)

R – S – R′

Ex:

h. The Nitrile Functional Group (cyano group)

$$R - CN$$

Ex.

5. MOLECULES WITH C=O FUNCTIONAL GROUPS

a. Introduction

- The $C=O$ group is called the **carbonyl group**.

b. General Structure

c. Aldehyde Functional Group

d. Ketone Functional Group

e. The Carboxylic Acid Functional Group

f. The Ester Functional Group

g. The Amide Functional Group

h. The Acid Chloride Functional Group

i. The Acid anhydride Functional Group

j. The Thioester Functional Group

6. HYDROCARBONS

   a. Alkenes
   b. Alkynes
   c. Arenes (Aromatics)

- See Tables _____ – _____ pages _____ – _____.

- Do Problems _____, pages _____ - _____

7. TYPES OF FUNCTIONAL GROUPS: A SUMMARY

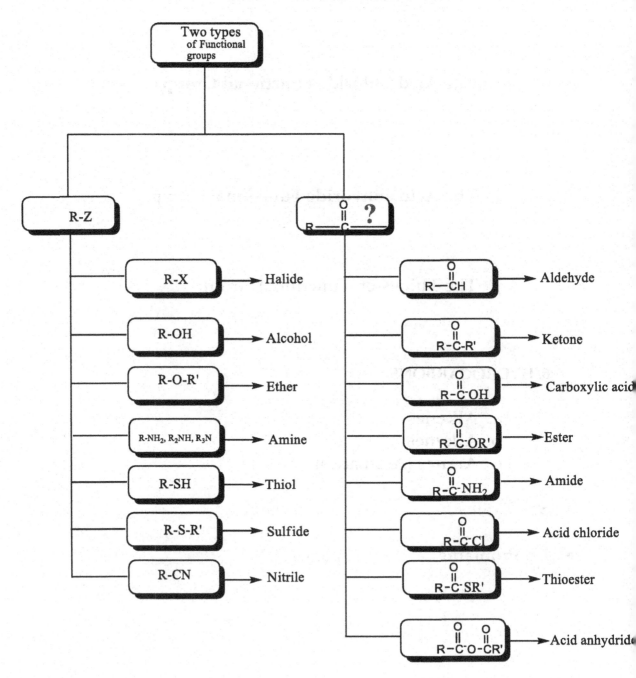

8. IMPORTANCE OF FUNCTIONAL GROUPS

- In OChem, functional groups are very important. The following characteristics of a given molecule depend on its functional group(s).
- Name of molecule
- Intermolecular forces
- Physical properties
- Molecular geometry
- Chemical properties (reactivity)

- Read pages _____ – _____. Do related problems.

C. PREDICTION OF REACTIVITY USING FUNCTIONAL GROUPS

1. INTRODUCTION

- Functional groups are usually the reactive sites on molecules.

Ex:

$$C{-}O$$

2. THE PRESENCE OF AN ELECTRONEGATIVE HETEROATOM IN A FUNCTIONAL GROUP

$$C{-}Z$$

Ex:

3. THE PRESENCE OF A LONE PAIR OF ELECTRONS ON A HETEROATOM IN A FUNCTIONAL GROUP = NUCLEOPHILIC SITE

- Note: a nucleophilic site is an "electron rich" site (an electronegative atom, a pi bond, etc...)

$$C{-}O{-}C \text{ or } C{-}S{-}C$$

- Read pages _____ - _____. Do problems on page _____

4. THE PRESENCE OF 1 OR 2 π BONDS = NUCLEOPHILIC SITE

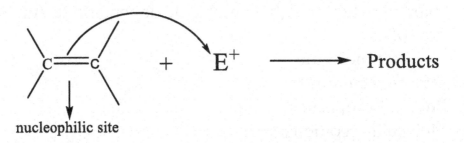

nucleophilic site

- Do Problem on page _____; Read page _____: Biomolecules (Structure of DNA).

5. GENERAL NUCLEOPHILE-ELECTROPHILE REACTIONS

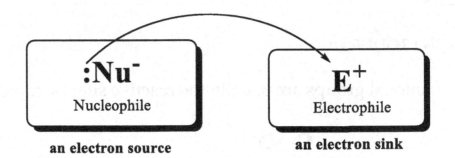

an electron source   an electron sink

D. CONCEPT OF CHEMICALLY EQUIVALENT HYDROGENS

1. TYPES OF CARBONS IN ALKANES

- Read pages _____ – _____.

$$R-\underset{\underset{H}{|}}{\overset{\overset{H}{|}}{C}}-H \qquad 1^\circ \text{ carbon}$$

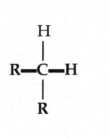

2° carbon

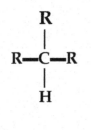

3° carbon

- Read page _____ and do problem_____.

2. EQUIVALENT HYDROGENS

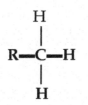

1° hydrogen

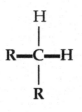

2° hydrogen

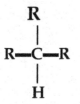

3° hydrogen

3. CLASSIFICATION OF AMINES

- Amines are classified in **4 ways**
- primary amines (1°)
- secondary amines (2°)
- tertiary amines (3°)
- quaternary amines = quaternary ammonium salts (4°)

- Read page _____.

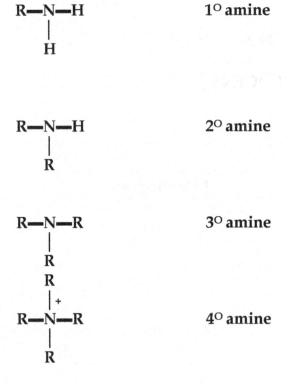

R—N—H
    |
    H
	1° amine

R—N—H
    |
    R
	2° amine

R—N—R
    |
    R
	3° amine

R—N—R (+)
    |
    R
	4° amine

- Read page _____ and do related problems_____.

E. INTERMOLECULAR FORCES: A REVIEW

1. DEFINITION

- Forces **between** molecules in a substance.

2. TYPES OF INTERMOLECULAR FORCES

- 3 types of IMF between organic molecules
- van der Waals forces (or LDF)
- Dipole - dipole interactions
- Hydrogen bonding

- See Table _____, page _____.

3. van der WAALS FORCES (VDW)

- **Weak forces** between all kinds of molecules (polar and nonpolar)
- Due to instantaneous dipoles created by change in electron distribution.

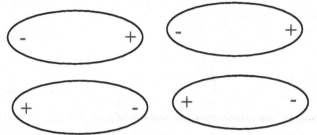

- Only forces existing between the molecules of **nonpolar substances.**
- Magnitude increases with molar mass due to increased polarizability.
- Magnitude also increases with increasing surface area.

- **Note: Hydrocarbons are nonpolar substances. Have only weak VDW between molecules.**

Ex: C_2H_6

- See Fig. _____ on page _____.

- See figure on page _____.

4. DIPOLE-DIPOLE INTERACTIONS (DD)

- Between molecules of polar substances

Ex: HCl

- See page _____.

5. H BONDING

- Between molecules of polar substances that contain **OH, NH,** and **HF.**

Ex:

6. OVERALL SUMMARY ON CHEMICAL BONDS

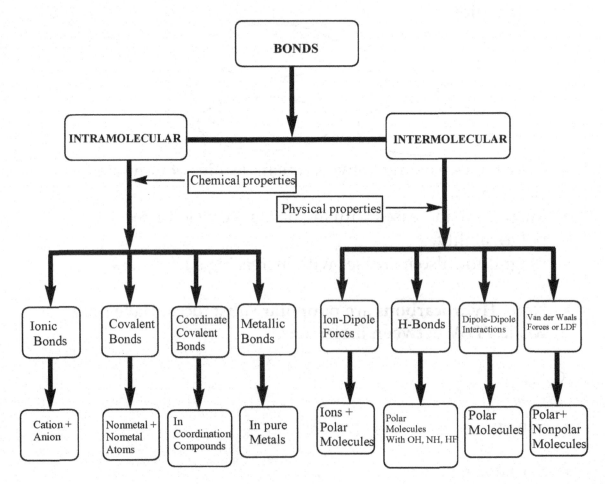

- Do problems on page _____.

- **Note: H bonds > DD > VDW.**

- **Note: In ionic compounds, the cation and anion are connected by ionic bonds.**

- See Table _____, page _____ for Summary.

E. INTERMOLECULAR FORCES AND PHYSICAL PROPERTIES

1. INTRODUCTION

- The solubility, boiling and melting points of a substance depend on the strength of its intermolecular forces.
- Weak IMF, low BP, low MP.
- Strong IMF, high BP, high MP.

2. BOILING POINT

a. Definition
- Temperature at which a liquid substance turns to gas at atmospheric pressure.

- BP depends on the strength of IMF, surface area, polarizability.

b. IMF
- Stronger IMF, high BP.

Ex: H$_2$O and CO$_2$

b. Surface Area

- Larger surface area, higher BP.

Larger surface area

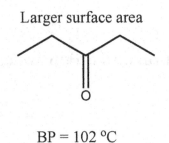

BP = 102 °C

Smaller surface area

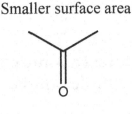

BP = 56 °C

- See Fig. _____, page _____.

c. Polarizability

- **Polarizability is the inability of an atom to hold its electrons tightly to itself or the ability of an atom to become polarized (or charged) or the ability of an atom to spread its electrons over a large surface area. A large, less electronegative iodine atom has a higher polarizability than the small electronegative F which can hold its electron more tightly because they are closer to its nucleus.**

- **Presence of polarizable atom (I), higher BP.**

I more polarizable than F

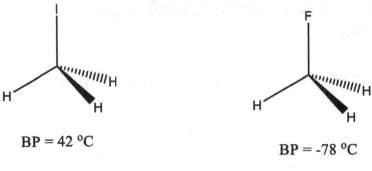

BP = 42 °C

BP = -78 °C

- See Fig. _____, page _____.

- Read pages _____ - _____.

- See example on page _____.

- Do problems on page _____.

3. MELTING POINT

a. Definition

- Temperature at which a solid substance turns to liquid at atmospheric pressure.

- MP depends on the strength of IMF and molecular symmetry.

b. IMF

- Stronger IMF, high MP.

Ex: Ice and dry ice

c. Molecular Symmetry

- For molecules with same functional group, **the most symmetrical has the highest MP**. Symmetry increases the ability of the substance to **pack into a crystalline structure**.

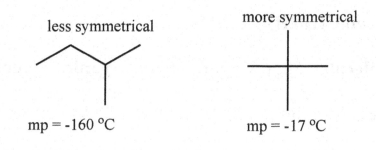

less symmetrical   more symmetrical

mp = -160 °C   mp = -17 °C

- See example on page _____.

- Do problems on page _____.

d. Straight-Chain Alkanes with Even and Odd Numbers of Carbon Atoms

- Experiments have shown that straight-chain alkanes with **even numbers** of carbon atoms have **higher** MP than those with **odd numbers** of C atoms. **Why?**

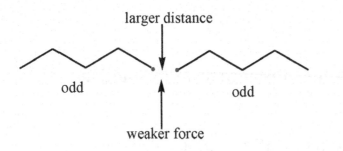 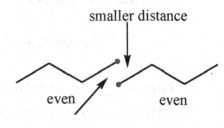

4. SOLUBILITY

a. Definition

- Amount of solute that can dissolve in a certain amount of solvent at a given temperature.

b. Water Soluble Compounds

- Ionic substances
- Organic compounds that contain **1 to 5 carbons and contain O or N atoms**.

Ex: CH_3CH_2OH and CH_3NH_2

c. Molecules That Are Soluble in Organic Solvents

- **All organic compounds.**

d. Like Dissolves Like

- **A polar solute** tends to dissolve in **polar solvents.**
- **A nonpolar solute** dissolves in **nonpolar solvents.**

Ex: I_2 in CCl_4; CH_3OH in H_2O

- **Note: Most ionic compounds are insoluble in organic solvents.**

- Read about Vitamins.
- Read about soaps and cell membranes.
- Read pages _____ - _____; Do Problems on pages _____ - _____.

- See Fig. _____, page _____.

- See Key Concepts, pages _____ - _____.

OCHEM UNIT 5: ALKANES AND CYCLOALKANES

A. GENERAL STRUCTURE OF THE ALKANES

1. DEFINITION

- **Alkanes**
- called aliphatic hydrocarbons
- also referred as saturated hydrocarbons (saturated with respect to H)
- contain only C—C single bonds (no double or triple bonds)
- contain **only** sp^3 hybridized carbons

Ex: C_3H_8

- Read page _____.

2. GENERAL MOLECULAR FORMULA OF ALKANES

$$C_nH_{2n+2}$$

- n = Total number of carbon atoms in molecule.
- Note: - We have a homologous series.

Ex:

- n = 1

- n = 2

- n = 6

- n = 20

- Read pages _____ - _____.
- Do Problem _____, page _____.

3. STRAIGHT CHAIN – BRANCHED ALKANES

a. Straight Chain Alkanes

- normal alkanes
- unbranched alkanes

Ex: C_4H_{10}

b. Branched Alkanes

- Have **carbon group(s)** (or **substituents**) attached to **parent chain (longest continuous carbon chain in molecule).**

Ex: C_8H_{18}

$$CH_3CH_2CHCH_2CH_2CH_2CH_3$$
$$|$$
$$CH_3$$

B. NAMING ALKANES

1. NOMENCLATURE

- Nomenclature is the art of naming chemical substances.

- **IUPAC** = International Union of Pure and Applied Chemistry, based in Geneva Switzerland.

- Sets rules for naming chemical compounds.

- # Name = Prefix -Parent – **Suffix**

Ex: C_4H_{10} but -an – e

- IUPAC: Names of all **saturated** hydrocarbons end in -ane, as in alk*anes*.

2. NAMING NORMAL (STRAIGHT-CHAIN) ALKANES

- Named according to the number of carbons in compound.

#carbons	prefix	name	molecular formula	condensed formula
1	meth	meth-an-e	CH_4	CH_4
2	eth			
3	prop			
	but			
	pent			
	hex			
	hept			
	oct			
	non			
	dec			
20	eicos			

- **Note: Please become familiar with all the prefixes!**

- **See Table _____, page _____.**

3. NAMING BRANCHED (OR SUBSTITUTED) ALKANES

a. Substituents or Groups

- A **substituent or a group is** a carbon group attached to the **main chain** or **parent chain (longest continuous carbon chain in the compound).**

Ex: $CH_3CH_2CH_2CHCH_3$
$$|$$
$$CH_3$$

b. Alkyl Groups

- **saturated** substituents that contain only **C and H atoms.**
- derive from the alkanes.
- names end in **–yl**
- General formula:

$$C_nH_{2n+1}-$$

Ex:

alkane	name	derived alkyl group	name
CH_4	methane	CH_3-	meth–yl
CH_3CH_3	Ethane	CH_3CH_2-	Eth–yl

c. Common Alkyl Groups

alkyl group	name	abbreviation
	methyl	Me
	ethyl	Et
	propyl	Pr
	isopropyl	i-Pr
	butyl	Bu
	isobutyl	i-Bu
	tert-butyl (tertiary butyl)	t-Bu
	sec-butyl (secondary butyl)	Sec-Bu
	neopentyl	n-Pent

- Read page _____.

d. Halogen Substituents

i. Halo Groups

element	name	group
F	fluor*ine*	fluor*o*
Cl	chlor*ine*	chlor*o*
Br	brom*ine*	brom*o*
I	iod*ine*	iod*o*

ii. Halide Groups

element	name	group
F	fluor*ine*	fluor*ide*
Cl	chlor*ine*	chlor*ide*
Br	brom*ine*	brom*ide*
I	iod*ine*	iod*ide*

Ex:

$$CH_3CH_2Cl$$

e. Rules for Naming Branched or Substituted Alkanes

- Read pages _____ - _____. Do all problems and examples.

i. Identification of the Parent or Main Chain

- Identify the parent chain = **the longest continuous carbon chain in compound.**

Ex:

$$CH_3CH_2CH_2CHCH_2CH_3$$
$$|$$
$$CH_3$$

ii. Numbering of the Parent Chain Carbons

- Number the carbons in the parent chain so that the first group encountered along the chain receives the **lowest possible number.**

Ex:

$$CH_3CHCH_2CHCH_2CH_3$$

$$CH_3 \qquad CH_3$$

iii. Locating the Groups

- Each group is then located by its name and the number of the carbon atom to which it is attached.

- If 2 or more identical groups are attached to the parent chain, then prefixes such as **di, tri, and tetra** are used.

Ex:

$$CH_3CHCH_2CHCH_2CH_3$$

$$CH_3 \qquad CH_3$$

iv. Alphabetical Listing

- If 2 or more types of substituents are present on the parent chain, list them **alphabetically**, but don't alphabetize by prefixes such as **di, tri, and tetra**.

Ex:

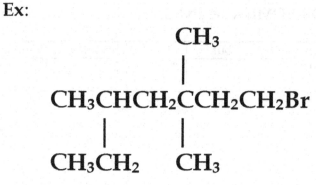

v. Alphabetizing of Some Groups

group	alphabetized under
isopropyl	i
ter-butyl	b
dimethyl	m

Ex: 4-ethyl - 3,3 - dimethylheptane

C. STRUCTURAL OR CONSTITUTIONAL ISOMERISM IN THE ALKANES

1. DEFINITION

- **Structural or constitutional** isomers are compounds that have the same numbers and kinds of atoms, but different atom connectivities.

- **Note: They have the same molecular formula, but different atom arrangements.**

Ex: C_6H_{14}

$$CH_3CH_2CH_2CHCH_3 \qquad CH_3CH_2CHCH_2CH_3$$
$$| \qquad\qquad\qquad |$$
$$CH_3 \qquad\qquad\qquad CH_3$$

2. NUMBER OF ISOMERS FOR SOME ALKANES

alkane	Methane	ethane	propane	but.	pent.	hexane	heptane
number of isomers	0	0	0	2	3	5	9

- See Table _____ , page _____ .

- Note: The number of possible isomers increases as the number of
  carbons increases. Some examples

- C_4H_{10}: 2 isomers

- C_5H_{12}: 3 isomers

- C_6H_{14}: 5 isomers

D. PROPERTIES OF THE ALKANES

1. PHYSICAL PROPERTIES

a. Introduction

- Alkanes are **nonpolar substances**. Therefore, they have only weak van der Waals forces (VWF or LDF) between their molecules.

b. Boiling Points of Alkanes

- **See Table _____, page _____.**

- Alkanes have low boiling points. The stronger the intermolecular forces, the higher the BP.

Ex:

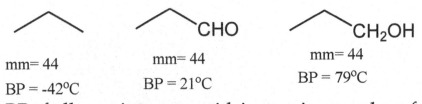

mm= 44

BP = -42°C

mm= 44

BP = 21°C

mm= 44

BP = 79°C

- BP of alkanes increases with increasing number of carbons. **Why?** Because of **increasing surface area**.

Ex:

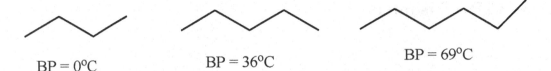

BP = 0°C

BP = 36°C

BP = 69°C

- **Note: Branched isomers have lower BP than straight chain alkanes. Why? Have decreased surface area in branched alkanes.**

Ex:

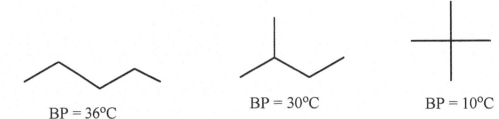

BP = 36°C

BP = 30°C

BP = 10°C

c. Melting Points of Alkanes

- Alkanes have low MP. The stronger the intermolecular forces, the higher the MP.

Ex.

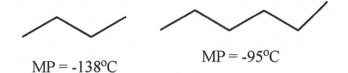

MP = -190°C   MP = -121°C

- In general, MP increases with increasing carbons (**increased surface area**).

Ex:

MP = -138°C   MP = -95°C

- **Recall: melting point increases with increased symmetry (See Unit 4).**

Ex:

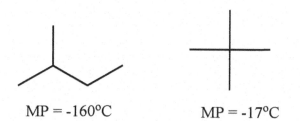

MP = -160°C   MP = -17°C

d. Solubility of Alkanes

- **Soluble in organic solvents.**
- **Insoluble in water.**

2. REACTIVITY OF ALKANES

- Alkanes **not very reactive**.
- **Parrafins**
- Undergo **halogenation** and **combustion** reactions.

E. CONCEPT OF CHEMICALLY EQUIVALENT HYDROGENS REVISITED

1. TYPES OF CARBONS

- **Read pages** _____ – _____.

1^O carbon

2^O carbon

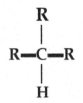

3° carbon

2. EQUIVALENT HYDROGENS

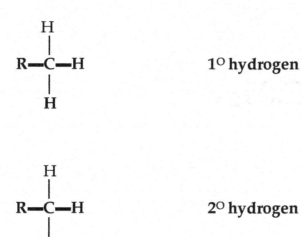

1° hydrogen

2° hydrogen

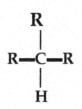

3° hydrogen

- Do example and all problems on page _____.

F. GENERAL STRUCTURE OF CYCLOALKANES

1. DEFINITION

- Saturated hydrocarbons that have at least one ring of carbon atoms.
- Alicyclic hydrocarbons (aliphatic and cyclic).
- Many found in nature.
- Ring size varies from 3 to 30 carbons.
- 5-member (**cyclopentanes**) and 6-member rings (**cyclohexanes**) are the most common.
- Rings of $-CH_2-$.

2. GENERAL MOLECULAR FORMULA

$$C_nH_{2n}$$

Ex:

- n = 3

- n = 4

- n = 5

- n = 6

- Do problems on page _____.

3. NAMING CYCLOALKANES

a. Naming Normal (Unsubstituted) Cycloalkanes

- In general

$$\boxed{\textbf{cyclo + alkane name}}$$

Ex:

- n = 3

- n = 4

- n = 5

- n = 6

b. Naming Branched (Substituted) Cycloalkanes

i. If have only one substituent on ring (monosubstituted ring), do not number. Count the number of carbons.

- If the number of carbons in ring is **lower** than the number of carbons in the group, name the compound as a **branched alkane. The ring becomes a group**.

- If the number of carbons in ring is **higher** than the number of carbons in the group, name the compound as a **branched cycloalkane.**
- If the number of carbons in ring is **the same** as the number of carbons in the group, name the compound as a **branched alkane or a branched cycloalkane.**

ii. Cycloalkyl Groups

Name	cyclopropyl	cyclobutyl	cyclopentyl	cyclohexyl
Structure				

Ex:

iii. If have two or more groups on ring, then number the substituents so that the second, or 3rd, or 4th group has **the lowest number**. Use **alphabetizing** as a guide.

Ex:

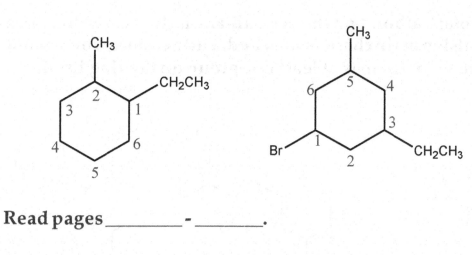

- Read pages _____ - _____.

- Read about common names, page _____.

4. NAMING COMPLEX CYCLOALKANES

- Cycloakanes do not always have "straightforward" structures as described earlier. Sometimes, they have **complex structures**. Indeed the outside chain itself can be branched.

Ex:

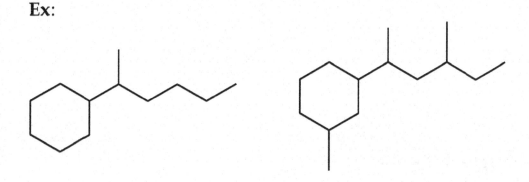

- They are named as:

(name of alkyl group on outside branch chain) (name of alkyl of outside main chain) name of cycloalkane

---

- **Note: Carbon 1 on the cycloalkane is the carbon to which the outside main chain is attached. Furthermore, you should list the "1" if there is at least one group on the ring itself.**

Ex:

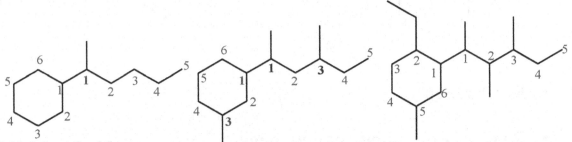

(1-Methylpentyl)cyclohexane

1-(1,3-Dimehylpentyl)-3-Methylcyclohexane

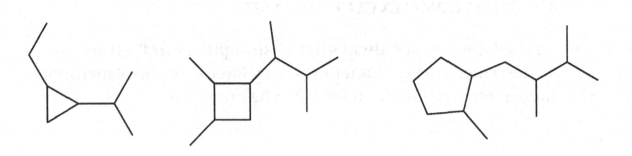

- **Read pages A-3 – A-4.**

100

5. NAMING BICYCLIC ALKANES

a. Types of Bicyclic Compounds or Two-Ring Compounds

- There are **3 kinds** of saturated **bicyclic** compounds:
- **Bridged Rings**
- **Fused Rings**
- **Spiro Compounds**

b. Bridged Rings

- In these types of compounds, the **two rings** share two **non-adjacent carbon atoms**.

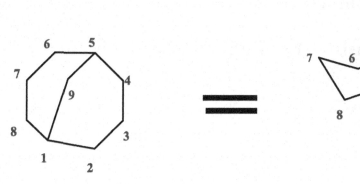   =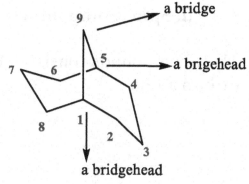

Ex:

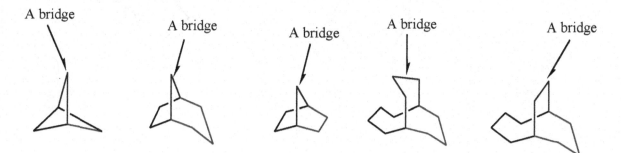

c. Fused Rings

- **A common carbon-carbon bond** is shared **by the two rings**

Ex:

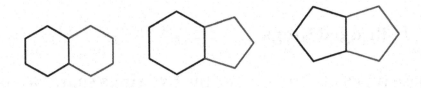

d. Spiro Compounds

- These compounds consist of **two rings** that share **only one carbon atom.**

Ex:

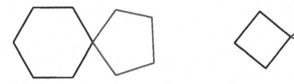

e. Naming Non-Substituted (Unbranched) Bridged and Fused Bicyclic Alkanes

bicyclo[#C in big ring. #C in small ring. #C in bridge]parent name of alkane

- **The parent name is based on the total number of carbons.**

Ex:

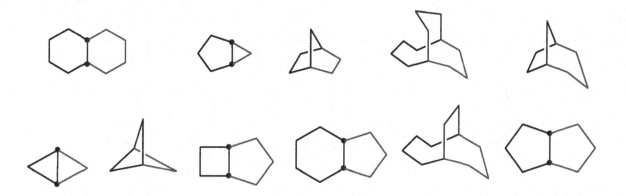

f. Naming Substituted (branched) Bridged and Fused Bicyclic Alkanes

- In this case, start numbering at a **bridgehead carbon**. Then, number the carbons of the **largest** and **smallest** rings in that order. Next, number the carbons in the bridge for **bridged compounds.**

Ex:

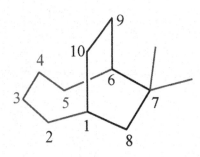

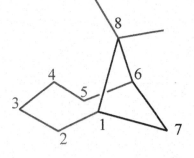

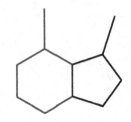

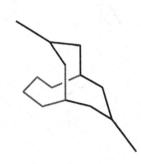

g. Naming Spiro Ring Compounds

- The carbons of **the smallest ring** are numbered **first**.

(substituent)-spiro[#C in small ring. #C in big ring]name of alkane

Ex:

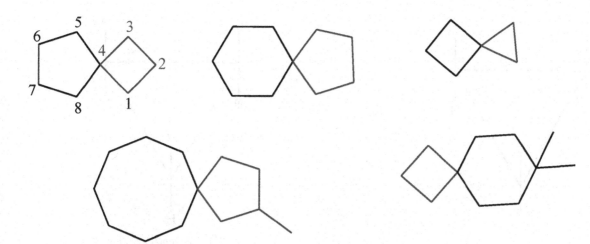

G. *cis-trans* ISOMERISM IN CYCLOALKANES

- **Stereoisomers** are compounds that have the same molecular formula, the same atom connectivities, but different spatial orientation of the atoms.

Ex: 1,2-dimethylcyclopropane

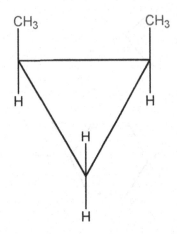

*cis-*1,2-Dimethylcyclopropane

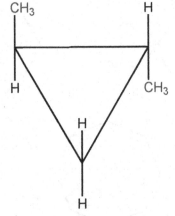

*trans-*1,2-Dimethylcyclopropane

H. SOURCES OF HYDROCARBONS: FOSSIL FUELS

- There are **3 main** sources:

    1. NATURAL GAS

- Natural gas is composed of **80% methane (CH$_4$), 10% ethane (C$_2$H$_6$), and 10% of other hydrocarbons.**

2. PETROLEUM

- Petroleum is a complex mixture of **hydrocarbons** that is believed to have derived from the **decomposition of marine organisms eons ago.**

- Read page _____.

- **Gasoline** = obtained from **fractional distillation.** The three major cuts in a refinery are:

- C_5H_{12} – $C_{12}H_{26}$ = gasoline
- $C_{12}H_{26}$ – $C_{16}H_{34}$ = kerosene (aviation fuel)
- $C_{15}H_{32}$ – $C_{18}H_{38}$ = Diesel fuel

Products of Fractional Distillation of Gasoline

fraction	boiling point ($^{\circ}$C)	use
C_1- C_5	-160 to 20	natural gas
C_5 - C_{12}	20 to 200	gasoline
C_{12} - C_{16}	175 to 275	kerosene
C_{15} - C_{18}	250 to 400	fuel oil
C_{18} - C_{25}	above 350	lubricants
above C_{25}	above 400	asphalt

- Read page _____.

- See figures on page _____.

- **Octane #** used against engine knock: **heptane** = worst fuel (O # = 0); **isooctane** = best fuel (O # = 100).

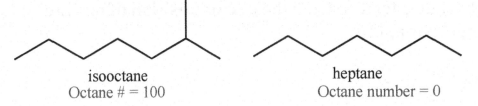

isooctane
Octane # = 100

heptane
Octane number = 0

3. COAL (CARBON)

- Hydrocarbons can be also obtained from **synthesis gas: CO + H_2.**

$$Coal + H_2O \longrightarrow CO + H_2 \xrightarrow{\text{catalyst}} \textbf{hydrocarbons}$$

I. REACTIONS OF ALKANES AND CYCLOALKANES

1. INTRODUCTION

- No functional groups in alkanes
- No π bonds
- Alkanes are very unreactive or inert.
- Undergo **combustion** and **halogenation** reactions = **redox reactions**

2. REDOX REACTIONS IN OCHEM

- **Oxidation = increase** in the number of **C-O** or **C-N** or **C-X** (X = halogens) bonds or **a decrease** in the number of **C-H** bonds.

Ex:

$$CH_4 + Cl_2 \longrightarrow CH_3Cl + HCl$$

$$CH_3CH_3 + Br_2 \longrightarrow CH_3CH_2Br + HBr$$

- Read pages _____ - _____. Do examples.

- **Reduction = decrease** in the number of C-O or C-N or C-X (X = halogens) bonds or **an increase** in the number of **C-H** bonds.

Ex:

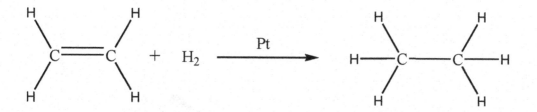

- Note: The more the C-O bonds, the more oxidized the compound.
  Ex: $CO_2 > HCO_2H > H_2CO$

3. COMBUSTION REACTIONS IN OCHEM (REDOX REACTIONS)

- The general combustion reaction for an **alkane** is

$$C_nH_{2n+2} + \frac{(3n+1)}{2}O_2 \longrightarrow nCO_2 + (n+1)H_2O$$

Ex:

- $C_5H_{12} + O_2 \longrightarrow$

- $C_nH_{2n} + O_2 \longrightarrow$

- $C_{15}H_{30} + O_2 \longrightarrow$

- **A Quick Look at Global Warming!**

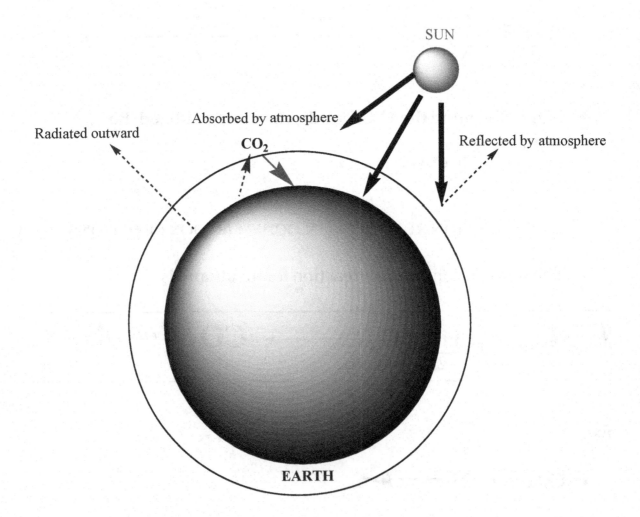

- Read about combustion of alkanes, global warming, and lipids.
- Read pages _____ - _____. Read about lipids.

4. HALOGENATION OF ALKANES (A REDOX REACTION): AN EXAMPLE

    a. Types of Hydrogens in Propane

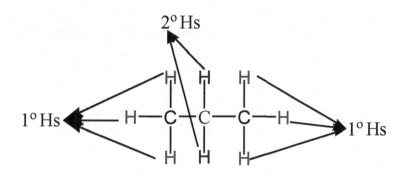

    b. Monochlorination of Propane: Two products

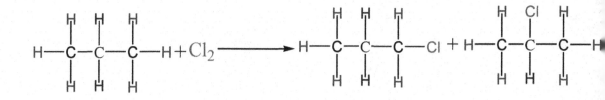

c. Dichlorination: 4 products

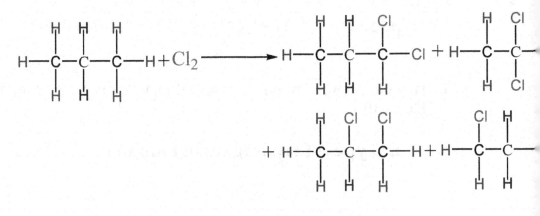

- **See Key Concepts, pages** _____ . _____

112

OCHEM UNIT 6: STEREOCHEMISTRY OF ALKANES AND CYCLOALKANES

A. INTRODUCTION

- **Stereochemistry** is the area of Chemistry concerned with the spatial arrangements of atoms in molecules.

- **Conformers** are isomers that can be interconverted by rotation about a single carbon-carbon bond.

B. CONFORMATIONS OF ETHANE

1. WAYS OF REPRESENTING ETHANE: 2

- Sawhorse.
- Newman projection.

a. Sawhorse Representation of Ethane

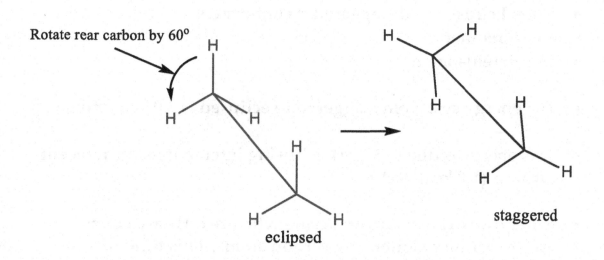

b. Newman Projection

i. Staggered and Eclipsed Conformations

- Take a look at the C—C axis. Have 2 conformations.

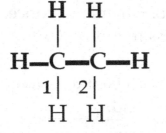

Staggered ——————▶ Eclipsed

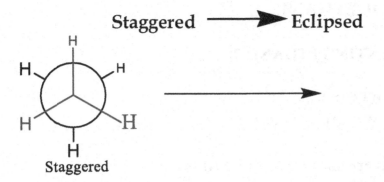

Staggered

- **Note: Eclipsed and staggered = conformers**
- **are stereomers**
- **are stereoisomers**

- The **energy cost** from **staggered to eclipsed is 3.0 kcal/mol**.

- This energy is due to **3 H-H eclipsing interactions on adjacent carbons (1.0 kcal each).**

- **Note: An H-H interaction occurs when two Hs are eclipsing each other on neighboring carbon atoms. This kind of interaction is called "torsional strain".**

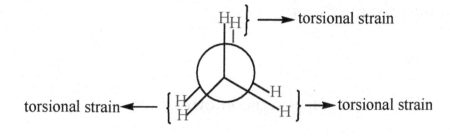

ii. Plot of Potential Energy vs. Dihedral Angle (Angle of Rotation)

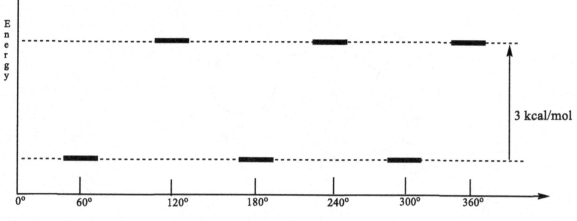

3 kcal/mol

Dihedral angle (angle of rotation)

- See figure on page _____.

- See Fig. _____, page _____.

- Read pages _____ - _____.

- Do Problems on pages _____ - _____.

C. CONFORMATIONS OF BUTANE

1. INTRODUCTION

- Have several eclipsed and staggered conformations.

- Take a look at C_2 — C_3 axis. See Fig. _____, page _____.

$CH_3CH_2CH_2CH_3$
1   2   3   4

Ex: Some examples of staggered conformations

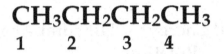

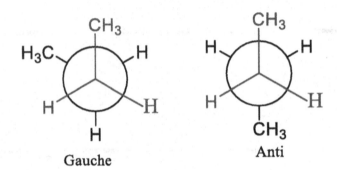

Gauche   Anti

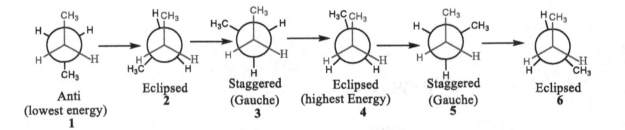

Anti (lowest energy) 1 → Eclipsed 2 → Staggered (Gauche) 3 → Eclipsed (highest Energy) 4 → Staggered (Gauche) 5 → Eclipsed 6

2. PLOT OF ENERGY vs. DIHEDRAL ANGLE

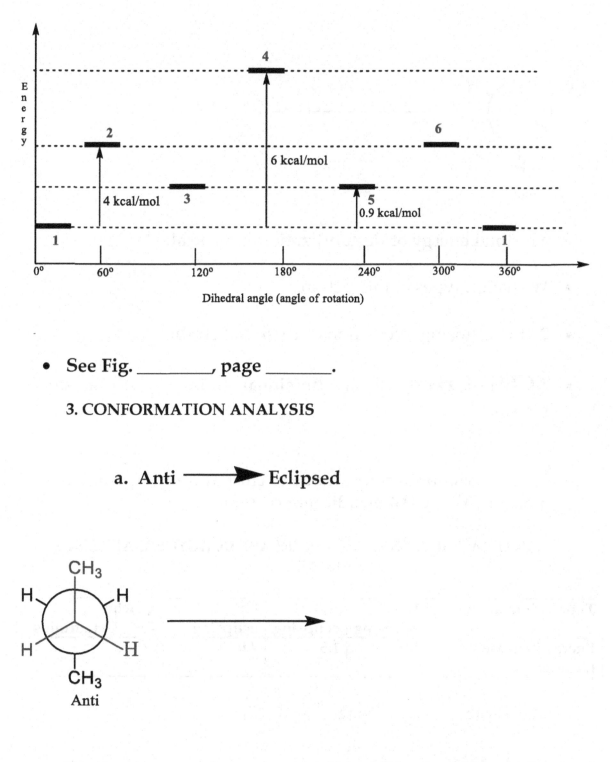

Dihedral angle (angle of rotation)

- See Fig. _____, page _____.

3. CONFORMATION ANALYSIS

a. Anti ⟶ Eclipsed

- Total energy cost (or **energy of destabilization**) = 1.0+1.5+1.5 = 4.0 kcal/mol = **torsional energy.**

117

b. Gauche ⟶ Eclipsed

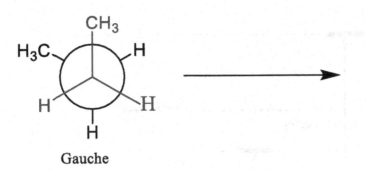

Gauche

- **The total energy of destabilization is 6.0 kcal.**

- We have 2 types of interactions:

- 2 H-H eclipsing interactions = **torsional strains**

- 1 **CH₃ - CH₃** = eclipsing (**torsional strain**) + repulsing (**steric strain**).

- **Note: Steric strain = strain due to closeness of big groups of atoms (CH₃ - CH₃) on adjacent carbons.**

3. INTERACTION ENERGIES INCREASE IN ACYCLIC ALKANES: A SUMMARY

Type of interaction	H, H eclipsing	H, CH₃ eclipsing	CH₃, CH₃ eclipsing	gauche CH₃, CH₃ groups
Energy increase (kcal/mol)	1.0	1.5	4.0	.9

- See Table _____, page _____.

- Read pages _____ - _____.

- Do problems on page _____.

D. STABILITY OF CYCLOALKANES

1. THE sp³ CARBON

- In cycloalkanes, C is sp³ hybridized and tetrahedral. The bond angles should be 109.5°.

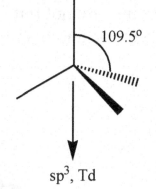

sp^3, Td

2. THE BAEYER ANGLE STRAIN THEORY (1885)

- Baeyer believed that cycloalkanes with ring size other than 5 or 6 carbons cannot exist because of **angle strain**, a strain induced by the deviation of a bond angle from the ideal tetrahedral value of 109.5°.
- **Note some examples below:**

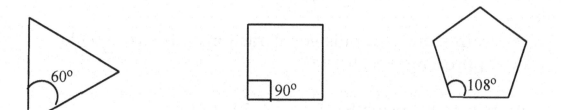

- Nowadays, cycloalkanes having 3 to 30 carbons (and beyond) have been prepared by chemists.
- **Conclusion**: Baeyer's theory is **wrong. Why?**
- Experiments have shown that:
- **cyclohexane is strain-free.**
- cyclopropane and cyclobutane are the **most strained.**

3. THE NATURE OF RING STRAINS

a. Puckered Conformations

- Baeyer's theory is wrong because he assumed that all cycloalkanes are flat. In reality, they are not flat. They just adopt puckered 3-dimensional conformations that have bond angles close to the ideal 109.5°.

Ex: Cyclobutane

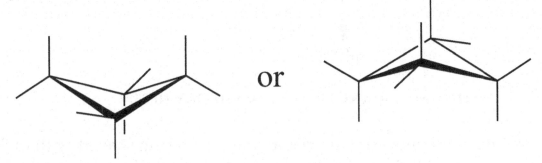

or

b. Factors Affecting the Stability of Cycloalkanes

- **3 factors**

- **Angle strain** = deviation of angle from 109.5°.
- **Torsional strain** = due to eclipsing of neighboring
- atoms.
- **Steric strain** = due to repulsive interactions   between bulky groups approaching each other.

Ex: Cyclopropane: 6 eclipsing Hs, 3 above, 3 below.

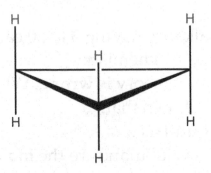

E. CONFORMATIONS OF CYCLOHEXANES

1. INTRODUCTION

- Widespread in nature.
- **Strain-free** (no torsional or angle strain). **Why?**
- Not flat
- Ring adopts a **chair conformation** that has C-C bond angles close to the ideal 109.5°.

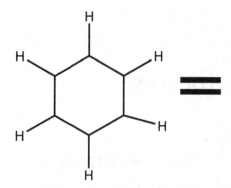

- **Note: All hydrogens are staggered; no eclipsed Hs, no torsional strain.**

2. AXIAL AND EQUATORIAL BONDS IN CYCLOHEXANE

- There are **2 types of bonds** in the chair conformation:

- 6 **axial** bonds = // ring axis.

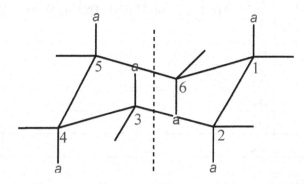

- 6 **equatorial** bonds = in rough plane of ring.

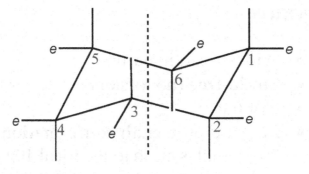

- See Fig. _____, page _____.

3. *cis-trans* POSITIONS IN CYCLOHEXANE

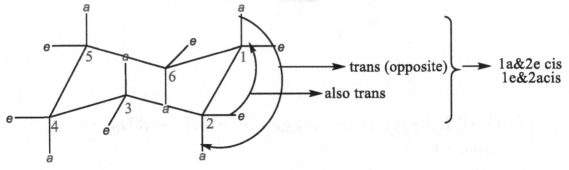

- Note: If *aa* of 2 carbons on the ring are trans implies that *ae* are cis; if *aa* are cis, then *ee* are also cis.

- Activity: Please, fill in the following table:

Group 1	Group 2	Cis-trans relationship
1a	2a	
2e	5e	
6a	4e	
2a	3e	
3a	5e	
4e	1e	

4. THE NEWMAN PROJECTION OF CYCLOHEXANE

- **Look through the C1-C2 and C5-C4 axes.**

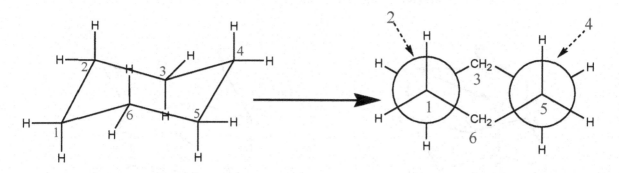

- **Note: All the Hs are staggered.**

5. RING FLIP IN MONOSUBSTITUTED CYCLOHEXANES

-The ring flips rapidly at room temperature.

Axial ⇌ equatorial

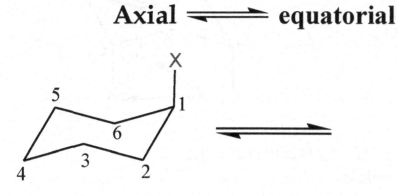

Ex:

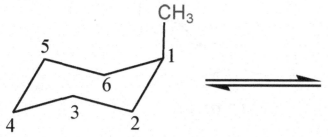

- **Note: Equatorial more stable than axial because of a lack of 1,3-diaxial interactions in the equatorial position (lack of steric strains between the X and the Hs).**

6. 1,3-DIAXIAL INTERACTIONS IN MONOSUBSTITUTED CYCLOHEXANES

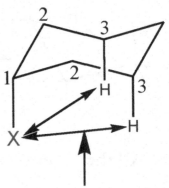

A 1,3-diaxial interaction   A 1,3-diaxial interaction

Ex:

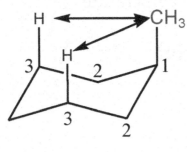

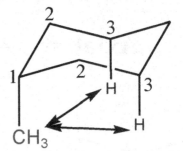

Ex: Which one is the most stable, cis and trans -1,2 - Dimethylcyclohexane? **Explain**

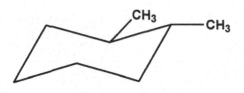

Trans -1,2-DMCH

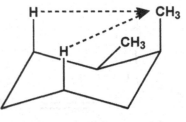

Cis -1,2-DMCH

7. BOAT CYCLOHEXANE

- Less stable.
- No angle strain, but does have **steric (close flagpole hydrogens)** and **torsional** strains due to **some** eclipsing Hs.

- **See figures, on page _____.**

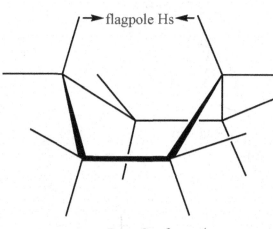

Boat Conformation

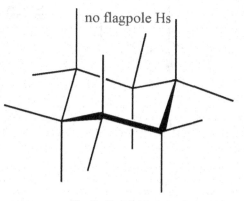

Chair Conformation

- **Read pages _____-_____.**

- **Do problems on page _____.**

F. CONFORMATIONS OF POLYCYCLIC MOLECULES

- A polycyclic molecule is made of 2 or more fused cyclohexane rings.

Ex: Decalin

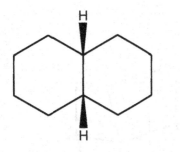

cis-decalin

=

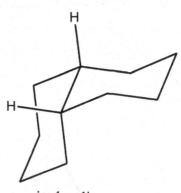

cis-decalin

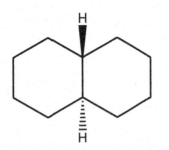

trans-decalin

=

trans-decalin

G. CONVERTING CYCLIC HEXAGONAL CYLOHEXANES TO CHAIR CONFORMATIONS AND VICE VERSA

1. UPS AND DOWNS IN CYCLOHEXANE

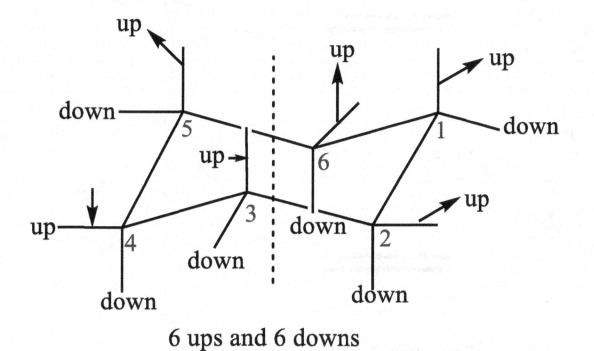

6 ups and 6 downs

2. TRANSFORMING CYCLIC CYCLOHEXANES TO THE CHAIR CONFORMATIONS AND VICE VERSA

- **Up in cyclic cyclohexane = Up in chair conformation**
- **Down in cyclic cyclohexane = down in chair conformation**

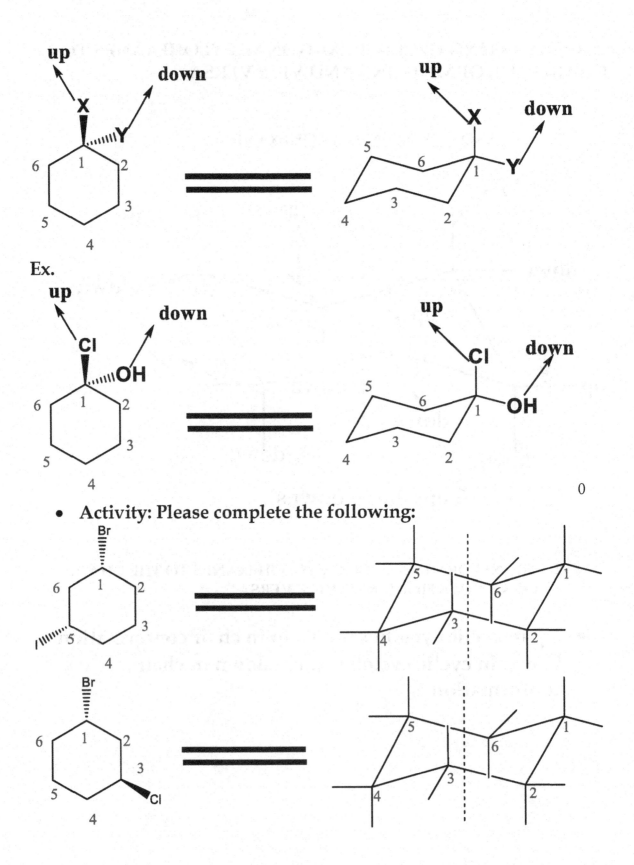

Ex.

• Activity: Please complete the following:

0

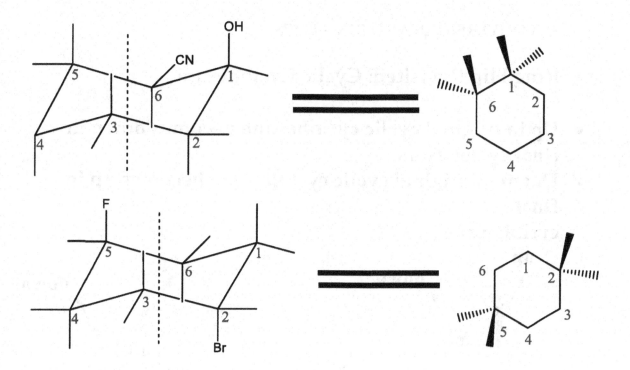

- **Ring Flip Revisited: Cyclic Cyclohexane**

- **Up in original cyclic cyclohexane becomes down in final cyclohexane.**
  - **Down in original cyclic cyclohexane becomes up in final cyclohexane**

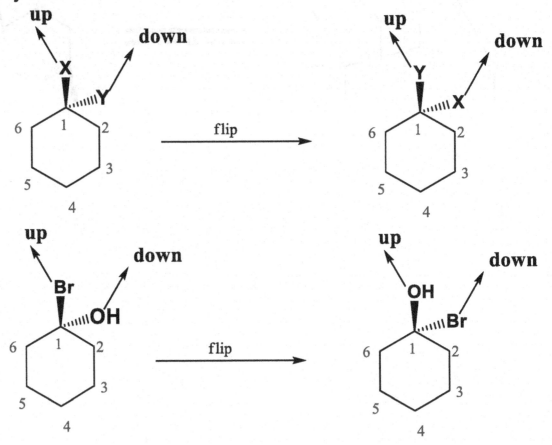

4. CONVERTING SAWHORSE PROJECTIONS TO NEWMAN PROJECTIONS AND VICE VERSA

- ### SAWHORSE TO NEWMAN PROJECTIONS

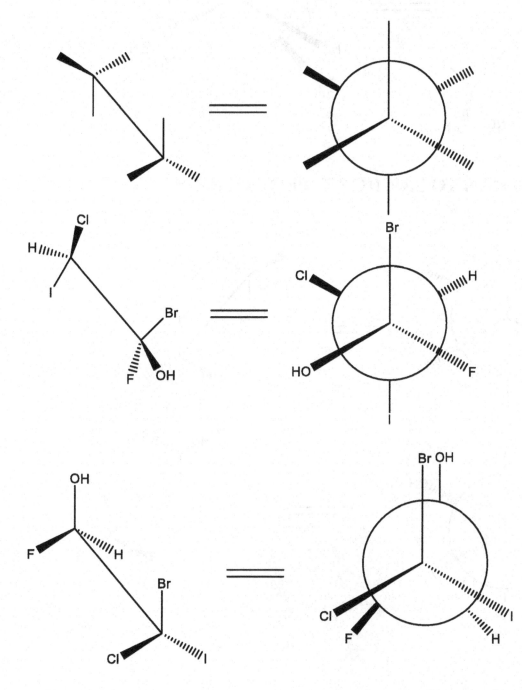

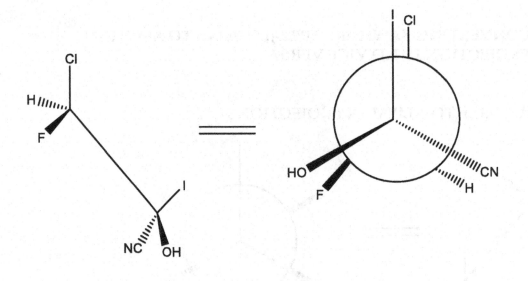

• NEWMAN TO SAWHORSE PROJECTIONS

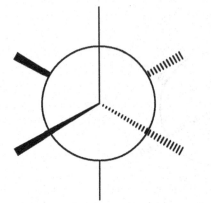 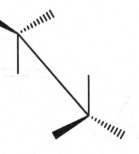

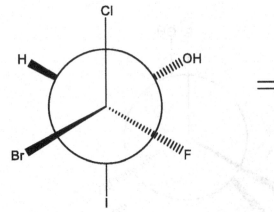 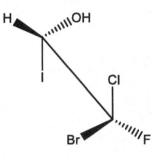

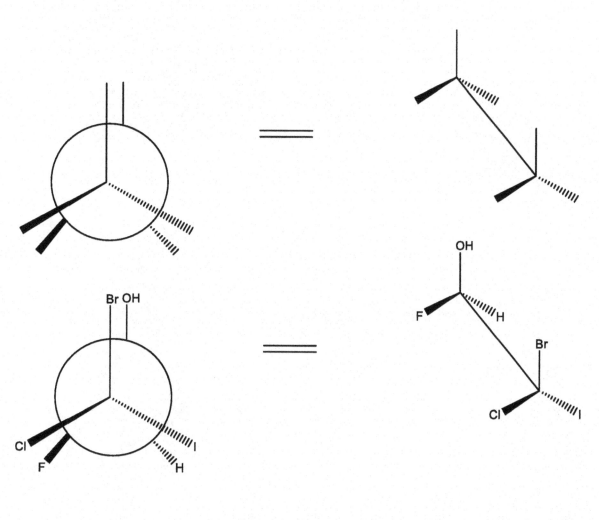

• **See Key Concepts, pages** _____ - _____.

OCHEM UNIT 7: STEREOCHEMISTRY: CHIRALITY

A. ISOMERISM REVISITED

1. INTRODUCTION

- There are two major types of isomers:
- Constitutional isomers.
- Stereoisomers.

2. STRUCTURAL OR CONSTITUTIONAL ISOMERS REVISITED

- Structural isomers are compounds that have the **same molecular formula**, but different atom connectivities. **They are not interconvertible by a simple C— C rotation.**

Ex: C_5H_{12}

$$CH_3(CH_2)_3CH_3 \text{ and } CH_3\overset{\overset{\displaystyle CH_3}{|}}{C}HCH_2CH_3$$

3. STEREOISOMERS OR STEREOMERS

- Stereoisomers are isomers that have the **same molecular formula, the same atom connectivities, and different spatial arrangements of their atoms.**

- There are 2 types of **configurations for stereoisomers:**

- **Enantiomers = mirror images**

- **Diastereomers = not mirror images**

Ex: *cis*-1,3-dimethylcyclopentane and *trans*-1,3-dimethylcyclopentane are **diastereomers.**

- Read pages _____ - _____ .

- Do Problem _____ page _____ .

B. HANDEDNESS AND CHIRALITY

1. HANDEDNESS

- Handedness = similarity to hands
- Left hand and right hand = not superimposable
- Relationship = mirror images

- A **chiral** object is an object that is **not superimposable** to its mirror image.

Ex: Left hand and right hand.

- A **chiral molecule** is a molecule that is **not superimposable** to its mirror image.

Ex: CBrClFI

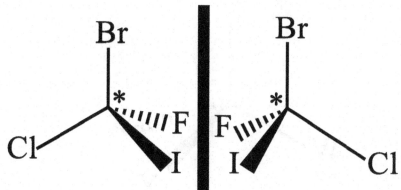

- **Enantiomers** are mirror images that are **not superimposable** to each other.

2. PLANE OF SYMMETRY

- A plane of symmetry is a plane that cuts an object into **equal halves**.

- **Note: Any object or molecule that possesses a plane of symmetry is superimposable to its mirror and is said to be achiral.**

- No enantiomers for an achiral molecule.

- **Read pages** _____ - _____.

- **Do problems on page** _____.

C. PREDICTION OF CHIRALITY

1. STEREOGENIC CENTER OR CHIRAL CENTER (ASYMMETRIC CARBON)

- A carbon that has **4 different groups** attached to it is **chiral** and is called a **(tetrahedral) stereogenic center (or asymmetric carbon)**.

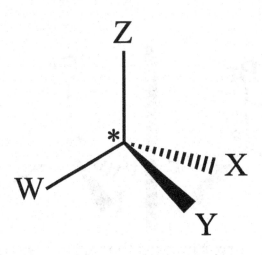

- **Note: A molecule that has 1 stereogenic center is always chiral.**

2. DISCOVERY OF CHIRALITY

- Discovered by French **Jean Baptist Biot** in 1815.
- Work extended by French **Louis Pasteur (1822-1895)**.

- **Note: Enantiomers have same connectivities, same MP and BP, same solubility in chiral solvents, same density, but are mirror images.** They are also affected differently by **plane-polarized light.**

3. PREDICTION OF CHIRALITY

    a. Write down the structural formula
    b. Identify the stereogenic center.
- If there is no stereogenic center, then the molecule is achiral.
- If there is at least 1 stereogenic center, but there is a plane of symmetry, then the molecule is achiral.
- If there is at least 1 stereogenic center and there is no plane of symmetry, then the molecule is chiral.

Ex: **CH₃C̲HClCH₂CH₃**

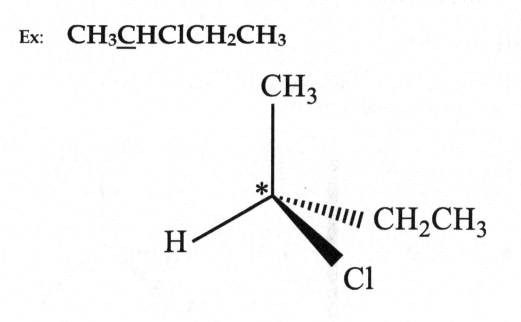

- Read pages _____ - _____.

- Do problems on pages _____ - _____.

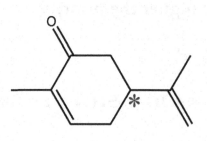

D. R-S CONFIGURATIONS

1. INTRODUCTION

- **R and S** configurations are used to differentiate between the 2 enantiomers of a chiral substance.

Ex:

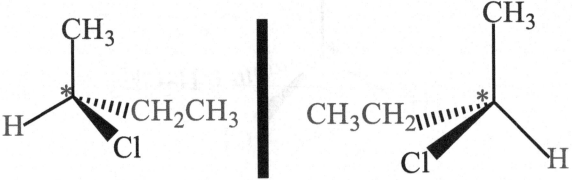

- **Note: R and S configurations are assigned using the Cahn-Ingold-Prelog priorities rules.**

2. GROUP PRIORITIES RULES

- **Rule #1:** Each atom directly attached to the stereogenic center is assigned a priority that is based on atomic number. **The higher the atomic number, the higher the priority.**

Ex: $I > Br > Cl > S > O > N > C > H$

- So: $-H < -CH_3 < -NH_3 < -OH < -SH < -Cl < -Br < -I$

- **Rule #2:** If a priority cannot be found for an atom directly attached to the stereogenic center on the basis of atomic number, **look at the next set of atoms and continue until a priority is found.**

Ex: $CH_3\underline{C}H(CH_2NH_2)CH_2CH_3$

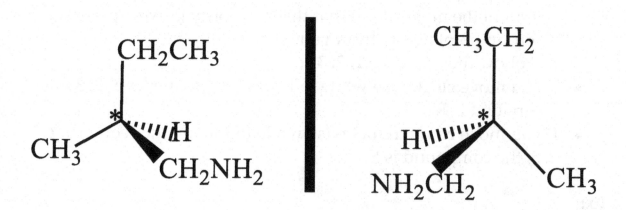

- **Rule # 3:** multiple bonded atoms are **considered** to be equivalent to the **same number of single bonded atoms.**

Ex:

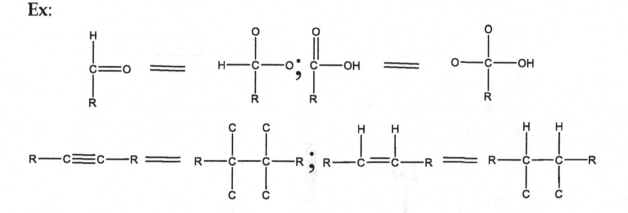

Rule #4: For isotopes bonded to a stereogenic center, priorities are assigned based on **decreasing mass number**.

Ex: T (H-3) > D (H-2)> H-1

- **Read** _____ - _____. **Do all problems.**

3. ASSIGNING R-S CONFIGURATIONS: molecular models recommended

    a. Assign a priority to each group about the stereogenic center.
    b. Orient the molecule so that the 4th priority (lowest-priority) group points away from you.
    c. Follow the sequence **1, 2, 3**.
- If you move **clockwise** while following the sequence **1, 2, 3**, the compound is **R**.
- If you move **counterclockwise** while following the sequence **1, 2, 3**, the compound is **S**.

Ex:

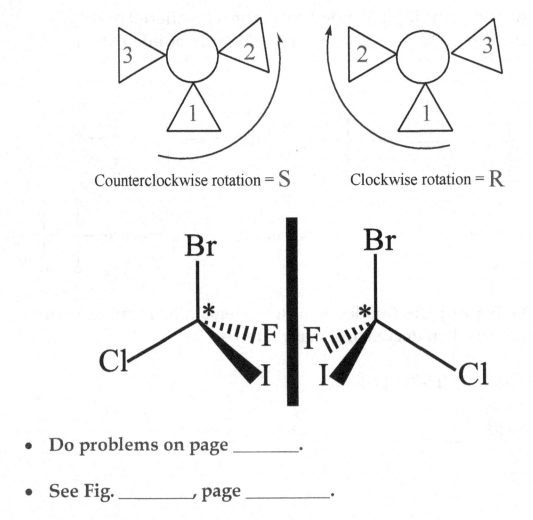

Counterclockwise rotation = S   Clockwise rotation = R

- Do problems on page _____.

- See Fig. _____, page _____.

4. ASSIGNING R-S CONFIGURATIONS WHEN THE LOWEST PRIORITY DOES NOT POINT AWAY FROM YOU

a. The Rotation Method

- The following 3 diagrams can help you assign R-S configurations when the assigned **lowest priority (4ᵗʰ group)** does not **automatically** point away from you.

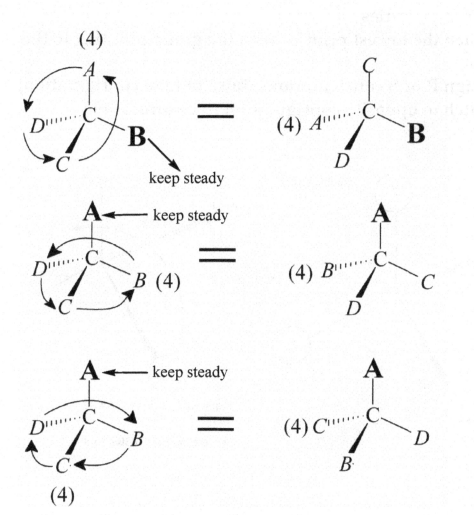

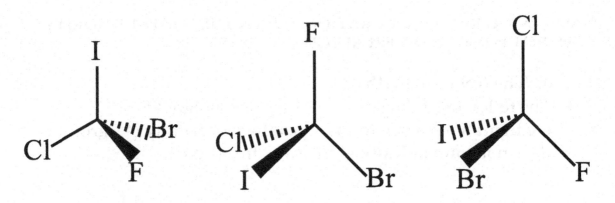

b. The Double Switch Method

- **Assign priorities**
- **Switch the lowest priority with the group pointing to the back**
- **Assign R or S configurations (false or fake configuration)**
- **Switch to opposite configuration (the correct one)**

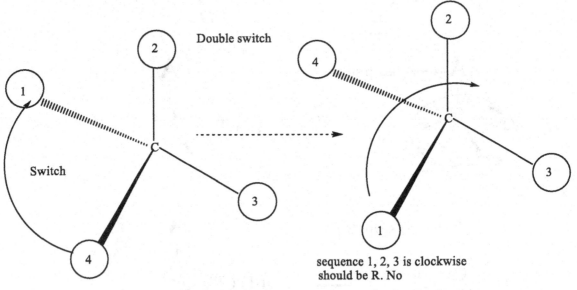

Double switch

Switch

sequence 1, 2, 3 is clockwise
should be R. No

Correct answer: switch to S

Ex: Use the double switch method to assign R/S configurations to the following chiral molecules.

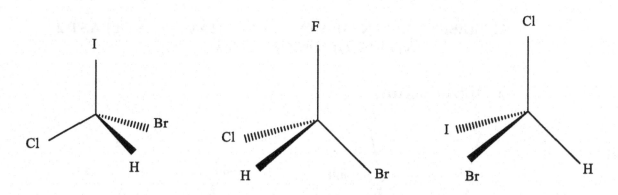

E. DIASTEREOMERS

1. DEFINITION

- **Diastereomers** are stereoisomers that are not mirror images.

- **Note: A compound with only 1 stereogenic center has 2 enantiomers.**

- **In general, a compound with n chiral centers has 2^n possible stereoisomers.**

Ex:

n = 1

n = 2

n = 3

n = 4

- **Do problems on page _____.**

Ex: $ClCH_2\underline{C}HClCl\underline{C}BrClCH_3$

2. REPRESENTATION OF MOLECULES HAVING AT LEAST 2 STEREOGENIC CENTERS

a. Introduction

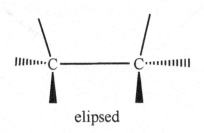
elipsed

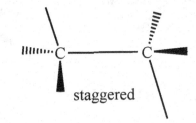
staggered

b. Drawing 2 eclipsed enantiomers A and B of $CH_3CHClCClBrF$ = 4 isomers

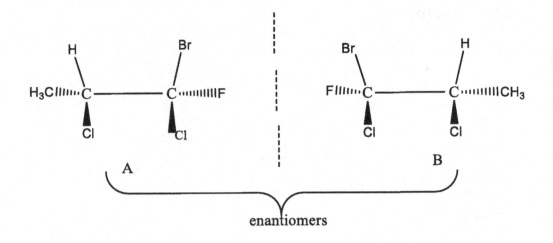

146

c. **Check if A and B are superimposable. Rotate B by 180° and compare the result, B′, to A**

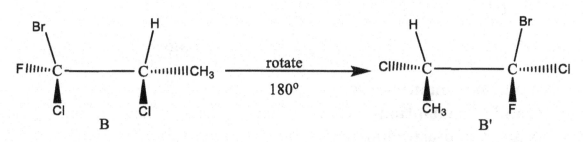

Conclusion: A different from B'

- **A and B = enantiomers**

   d. **Finding the Other 2 Isomers: Switch 2 Groups in A: CH₃ and Cl**

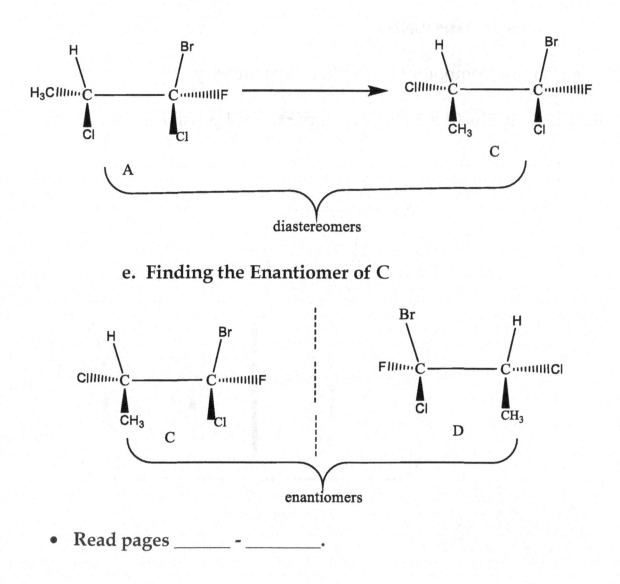

e. **Finding the Enantiomer of C**

- **Read pages _____ - _____.**

- Do problems on pages _____, _____, _____, _____, and page _____.

- A Summary: See Figs. _____, _____ pages _____ - ___.

- A and B = enantiomers
- C and D = enantiomers
- A and C = diastereomers
- A and D = diastereomers
- B and C = _____
- B and D = _____

3. MESO COMPOUNDS

- A **meso** compound has a plane of symmetry.

Ex: ClBr$\underline{C}$H$\underline{C}$HBrCl = 2 chiral centers ➜ 2^2 = 4 possible isomers.

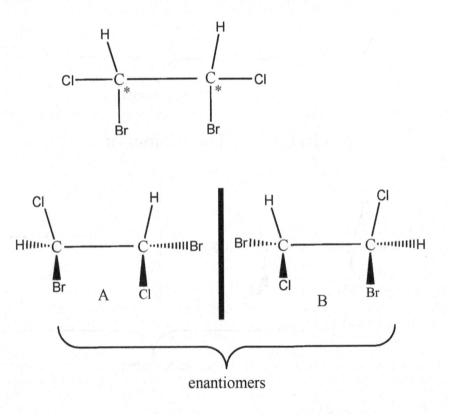

enantiomers

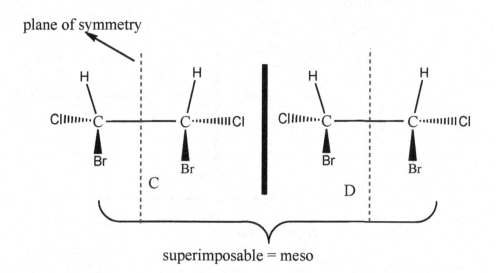

plane of symmetry

C

D

superimposable = meso

- **Note: If a compound has n stereogenic centers and a plane of symmetry, then it has $2^n - 1$ stereoisomers, including one meso compound.**

- **Note: The meso and the 2 enantiomers have different chemical and physical properties. The meso is achiral since it has a plane of symmetry.**

F. ASSIGNING R – S CONFIGURATIONS TO COMPLEX MOLECULES

1. USING NEWMAN PROJECTIONS: Eclipsed

Ex: Tartaric acid

$$HO_2C - \overset{\overset{\displaystyle H}{|}}{C^*} - \overset{\overset{\displaystyle H}{|}}{\underset{\underset{\displaystyle OH}{|}}{C^*}} - CO_2H$$
$$\qquad\quad \underset{OH}{}$$

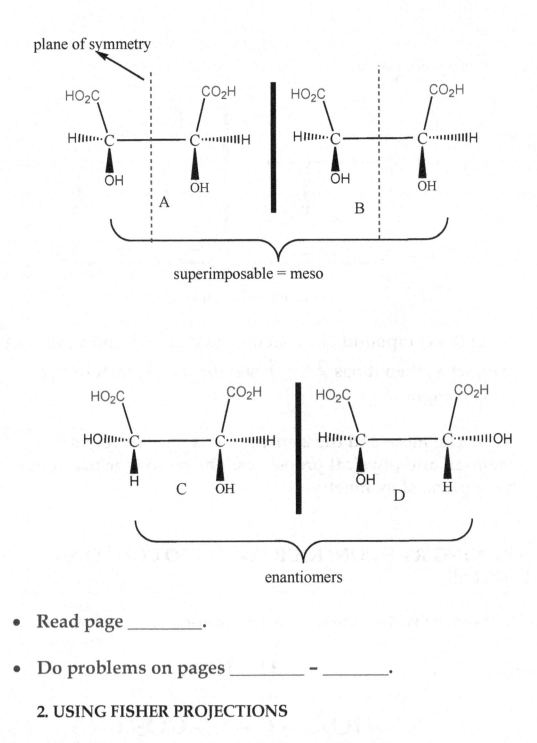

plane of symmetry

A   B

superimposable = meso

C   D

enantiomers

- Read page _____.

- Do problems on pages _____ – _____.

2. USING FISHER PROJECTIONS

- **Vertical lines = bonds going into the back**
- **Horizontal wings = bonds coming towards us**
- **Carbon skeleton curls around a barrel. See Fig. _____, page _____.**

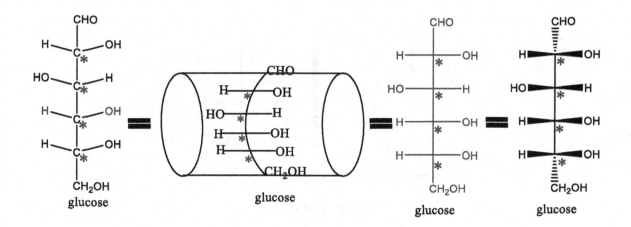

glucose = glucose = glucose = glucose

Ex: Tartaric acid

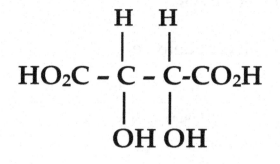

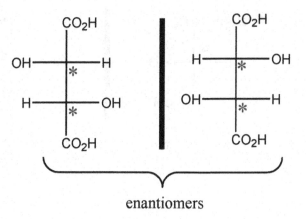

enantiomers

151

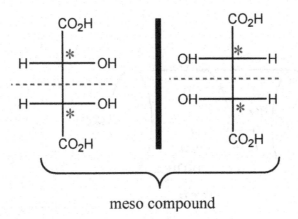

meso compound

3. ASSIGNING D and L CONFIGURATION TO FISHER PROJECTIONS

- **Reference**: Glyceraldehyde

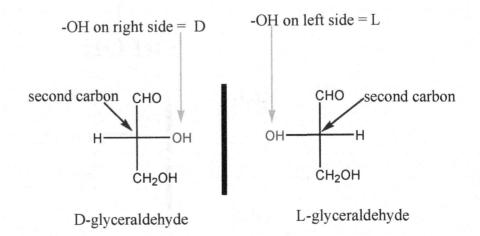

-OH on right side = D

-OH on left side = L

second carbon   CHO

H——OH

CH₂OH

D-glyceraldehyde

CHO   second carbon

OH——H

CH₂OH

L-glyceraldehyde

- **Note: If the –OH group on the second carbon from the bottom is on the right hand side, then we have a D. Otherwise, we have an L. See some C₆ examples below.**

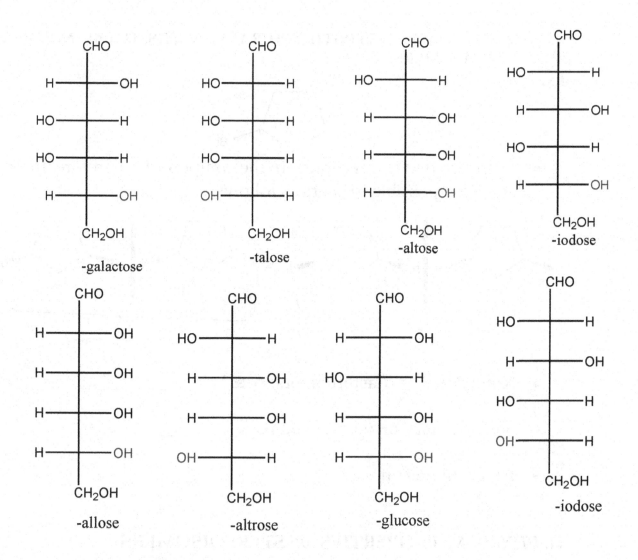

-galactose

-talose

-altose

-iodose

-allose

-altrose

-glucose

-iodose

G. STEREOISOMERISM IN DISUBSTITUTED CYCLOALKANES

1. INTRODUCTION

- There are 2 ways of looking at structures:
- Total number of isomers based on the **number of stereogenic centers.**
- cis-trans isomers.

2. CYCLOALKANES WITH 2 CHIRAL CENTERS: TOTAL NUMBER OF ISOMERS

Ex:

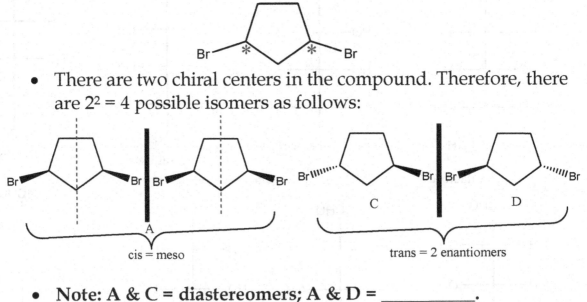

- There are two chiral centers in the compound. Therefore, there are $2^2 = 4$ possible isomers as follows:

- **Note: A & C = diastereomers; A & D = _____.**

- **See summary on isomers on page _____.**

- Do problems on pages _____ – _____.

H. PHYSICAL PROPERTIES OF STEREOISOMERS

1. POLARIMETRY

- **Polarimetry** is an experimental method used to detect chirality.

2. PLANE-POLARIZED LIGHT

- **Plane-polarized light** is light that vibrates in only one **specific plane.**

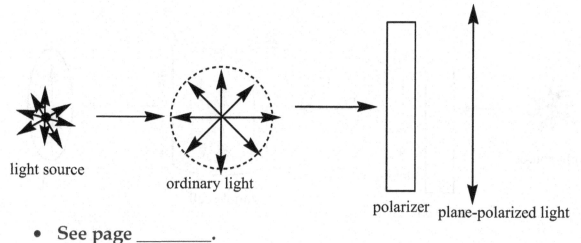

light source

ordinary light

polarizer plane-polarized light

- See page _____ .

3. THE POLARIMETER

- A **polarimeter** is a device used to detect **optical activity.** See figure on page _____ .

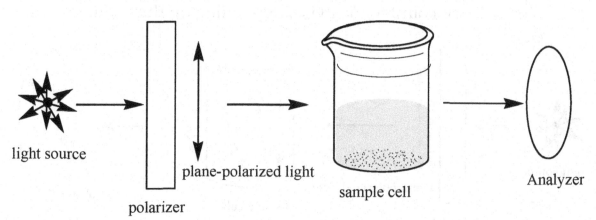

light source

plane-polarized light

polarizer

sample cell

Analyzer

4. BEHAVIOR OF SUBSTANCES INSIDE A POLARIMETER

a. Achiral Substances in a Polarimeter

- Achiral substances **do not** rotate the plane of polarized light; They are said to be **optically inactive.** When a sample of an achiral substance is in place, the angle of rotation is **zero (0)** since the plane of polarized light is **unchanged. See figure on page** _____ .

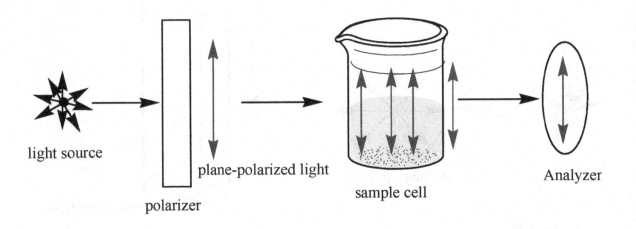

light source

plane-polarized light

polarizer

sample cell

Analyzer

b. Chiral Substances in a Polarimeter

- Chiral substances rotate the plane of polarized light either clockwise or counterclockwise, depending on the enantiomer in place.

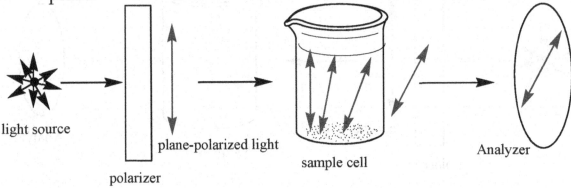

light source

plane-polarized light

polarizer

sample cell

Analyzer

- The enantiomer that rotates the plane of polarized light **clockwise** is said to be **dextrorotatory.** It is called the *d* or the (**+**) isomer.
- The other enantiomer which rotates the plane of polarized light **counterclockwise** is said to be **levorotatory.** It is called the *l* or the (**-**) isomer.
- Either way, the substance is said to be **optically active**.

- **Note: Meso isomers have 0 rotation; they are achiral.**

5. RACEMIC MIXTURES OR RACEMATES

- A **racemate** is a mixture that contains **equal amounts** of a pair of enantiomers **(50:50 mixture)**

- Note: The process of mixing equal amounts of two enantiomers is called racemization.

- Note: A racemate is optically inactive since its angle of rotation in a polarimeter is zero.

- See Table _____, page _____.

- Do Problems, pages _____ - _____.

- Read pages _____ - _____.

6. SPECIFIC ROTATION: $[\alpha]_D$

$$[\alpha]_D = \frac{\text{observed rotation}}{\text{(pathlength)} \times \text{conc (g/mL)}}$$

or

$$[\alpha]_D = \frac{\alpha \text{ observed}}{l \times C}$$

or

$$\alpha_{\text{observed}} = [\alpha]_D \, l \times C$$

- where the measurement of α_{observed} is performed at the λ of the **yellow line of Na (D line, 589 nm):**

- l = pathlength of sample cell in decimeter
- C = concentration in g/mL

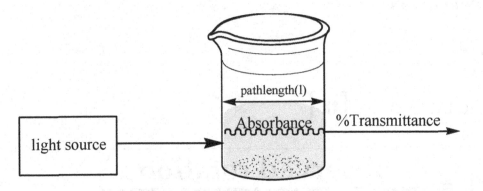

- **Note: By convention**

  - $[\alpha]_D = +$ for the *d* enantiomer

  - $[\alpha]_D = -$ for the *l* enantiomer

Ex: For (+) sucrose, $[\alpha]_D = + 66.47$

For (-) sucrose, $[\alpha]_D = - 66.47$

158

- Note: $[\alpha]_D$ is a constant and an intrinsic property (like density, MP, BP, etc.). It is the rotation obtained from a 1.0-g/mL sample solution observed in a 1.0-dm polarimeter at λ = 589 nm (sodium D-line) and a temperature of 25^{0}C. It is independent of concentration. On the other hand, α observed is a function of concentration, temperature, wavelength, and pathlength. For instance, if you double the concentration of the substance, α observed will also double. Likewise, if you halve the concentration, α observed will be halved.

Ex 1: When 3.00 g of a substance is dissolved in 20.0 mL of acetone, the observed rotation was -4.31^o. What is the specific rotation ($[\alpha]_D$) of this compound? (l = 6.00 cm).

$$[\alpha]_D = \frac{\alpha \text{ observed}}{l \times C}$$

- l = 6.00 cm = 6.00/10 dm = .600 dm

- C = 3.00 g/20.0 mL = .150 g/mL

$$[\alpha]_D = \frac{-4.31^o}{.600 \text{ dm} \times .150 \text{ g/mL}}$$

$$= -47.9^o$$

Ex 2: The molarity of a chiral substance (molar mass 205 g/mol; $[\alpha]_D = +36.5$) is .250 M. Calculate the observed rotation in a polarimeter in which the pathlength is 10.0 cm.

- Note: For a racemate, $[\alpha]_D = 0$.

- Do Problem _____, page _____.

7. αobserved FOR A NON-RACEMIC MIXTURE OF R AND S (OR D and L)

$$\alpha_{obs(mixt)} = \alpha_{obs}(R) + \alpha_{obs}(S)$$

or

$$\alpha_{obs(mixt)} = \left[\left([\alpha]_{D(R)} \, l \, C_R\right) + \left([\alpha]_{D(S)} \, l \, C_S\right) \right]$$

or

$$\alpha_{obs(mixt)} = \alpha_{obs}(D) + \alpha_{obs}(L)$$

or

$$\alpha_{obs(mixt)} = \left[\left([\alpha]_{D(D)} \, l \, C_D\right) + \left([\alpha]_{D(L)} \, l \, C_L\right) \right]$$

Ex: Calculate the expected observed rotation of the mixture of the enantiomers of 2-pentanol ($C_5H_{12}O$; molar mass = 88.1 g/mol) when equal volumes of a 1.50-M solution of R and a .950-M solution of S are mixed in a 10.0-cm sample tube. ($[\alpha]^D$ (S) = 10.50° mL.g^{-1}dm^{-1}.)

160

- **Answer:**

$([\alpha]^D$ (R) $= -10.50°$ mL.g^{-1}dm^{-1}

$l = 10.0$ cm $= 1.00$ dm

- Calculate C_R and C_s in the solution. Assume V is the initial common volume; the total volume of the mixture is 2V

C_R x 2V = 1.50 x V => C_R = (1.50 x V)/(2V) = 1.50 x (1/2) = .750 M

C_R = .750 mol/1L x (88.1g/1 mol)x (1L/1000 mL)

$$= 6.61 \times 10^{-2} \text{ g/mL}$$

C_S x 2V =.950 x V => C_S = (.950 x V)/(2V) = .950 x (1/2) = .475 M

C_S = .475 mol/1L x (88.1g/1 mol)x (1L/1000 mL) = 4.18 x 10^{-2} g/mL

$$\boxed{\alpha_{\text{obs(mixt)}} = \left[\left([\alpha]_{D(R)} \, l \, C_R\right) + \left([\alpha]_{D(S)} \, l \, C_S\right)\right]}$$

= [(-10.50)x((1.00)x(6.61 x 10^{-2})] + [(+10.50)x((1.00)x(4.18 x 10^{-2})]

= -.694 + .439

=-.255°

8. **USING % OF R AND S (or D and L) TO GET $[\alpha]_{\text{obs(mixt)}}$**

$$\boxed{\alpha_{\text{obs(mixt)}} = \left[\left(\%[\alpha]_{D(R)} \, l \, C_R\right) + \left(\%[\alpha]_{D(S)} \, l \, C_S\right)\right]}$$

or

$$\boxed{\alpha_{\text{obs(mixt)}} = \left[\left(\%\alpha_{\text{obs}(R)}\right) + \left(\%\alpha_{\text{obs}(S)}\right)\right]}$$

Ex: Calculate the observed rotation of a mixture of 75.0%R and 25.0% S. (α_{observed} (R) = -115.0° mL.g^{-1}dm^{-1}).

- Answer:

- $\alpha_{obs(mixt)}$ = .750 x (-115.0) + .250 x (+115.0) = -86.3 + 28.8 = -57.6°

9. ENANTIOMETRIC EXCESS (OR OPTICAL PURITY)

a. Introduction

- Ordinarily, pure enantiomers and pure racemates are rare. In reality, we usually have mixtures made of unequal amounts of both enantiomers of a given substance. Suppose a certain chiral substance has two isomers A and B. In a certain mixture of A and B, either one of these two enantiomers could be in excess. Let's assume that **A is in excess.** We can calculate the **enantiometric excess (or optical purity), ee,** as follows:

$$\boxed{ee = \%A - \%B}$$

Ex: A mixture is made of 70% A and 30% B. What is the ee?

$$ee = 70 - 30 = 40\%$$

b. Calculation of the % for Each Isomer from the ee

- Suppose the ee of a mixture of A (in excess) and B is known. From this information, the respective %A and %B can be calculated.

$$\boxed{\%A = ee + \frac{(100 - ee)}{2}}$$

or

$$\boxed{\%A = (1/2)ee + 50}$$

and

$$\boxed{\%B = 100 - \%A}$$

Ex 1: The ee of a mixture of A and B is 80%. Calculate the % of each enantiomer present.

ee= 80% means there is 80% excess of A in the non racemic mixture.

$$\boxed{\%A = 80 + \frac{(100 - 80)}{2}}$$

$$= 90\%$$

and

$$\boxed{\%B = 100 - 90}$$

$$= 10\%$$

Ex 2: The ee of a mixture of A (in excess) and B is 35%. Calculate the respective % of A and B.

- Do problems on page _____.

  c. Getting the ee from $[\alpha]_D$

- **ee** can also be calculated from $[\alpha]_D$.

$$\boxed{ee = 100 \times \frac{[\alpha]_{\text{mixture}}}{[\alpha]_{\text{pure enantiomer}}}}$$

163

Ex: $[\alpha]_D$mixture = -10; $[\alpha]_D$ pure = -40. What is ee? Calculate the %A and %B.

- Do example on page _____ .

8. ENANTIOMERS VS. DIASTEREOMERS

- **Enantiomers :** Mirror images that have the same MP, BP, density. They are very difficult to separate by **physical means**.

- **Diastereomers:** Not mirror images; these can be separated.

Ex: Tartaric acid
- See Fig. _____ , page _____ .

- Do _____ , page _____ .

I. CHIRALITY IN THE BIOLOGICAL WORLD

- A chiral substance reacts only with a molecule that matches its **own chirality**. In other words, **l reacts with l and d reacts with d**. This is why enzymes are so specific.
- This is applied in drug design.

Ex: Your left shoe.

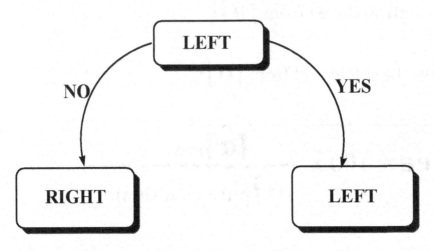

- Read about Ibuprofen, Advil, and Carvone pages _____ - ____.

- **Note: Chiral recognition is the phenomenon in nature in which a chiral receptor interacts differently with different enantiomers.**
- The following examples illustrate **chiral recognition**.
- See Fig. _____, page _____.

Ex 1: R and S carvone

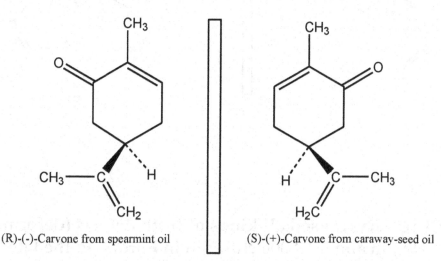

(R)-(-)-Carvone from spearmint oil   (S)-(+)-Carvone from caraway-seed oil

- **Odors: According to the Amoore Stereochemical Theory of Odor, if the receptor sites of the nose are chiral, then they will respond differently to different enantiomers. Whereas, the R carvone has a pleasant odor, its S-enantiomer stinks. Why?**
- **Toxicity: The S is 400 times more toxic to rats than the R counterpart.**

- **Activity: $[\alpha]_D$ (R-carvone) = -62.5°. What is $[\alpha]_D$ (S)?**

- See page _____.

Ex 2: R and S Thalidomide

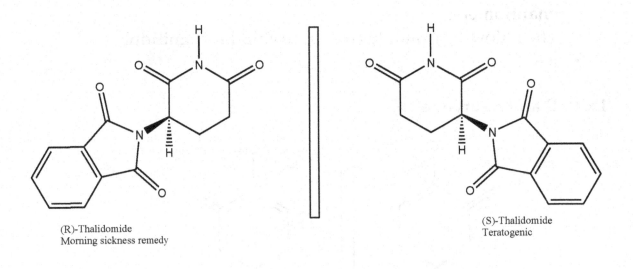

(R)-Thalidomide
Morning sickness remedy

(S)-Thalidomide
Teratogenic

- **Note: The S has caused all kinds of birth defects (deformed limbs, etc.) in about 10,000 children in Europe in the late 50s.**

Ex 3: R and S (Advil and Motrin) Ibuprofen

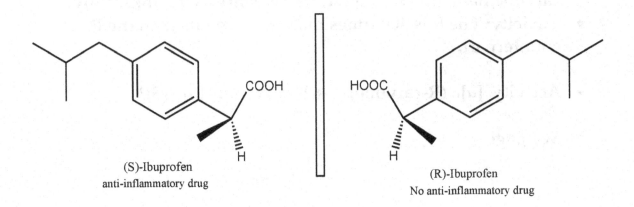

(S)-Ibuprofen
anti-inflammatory drug

(R)-Ibuprofen
No anti-inflammatory drug

Ex 4: R and S Naproxen

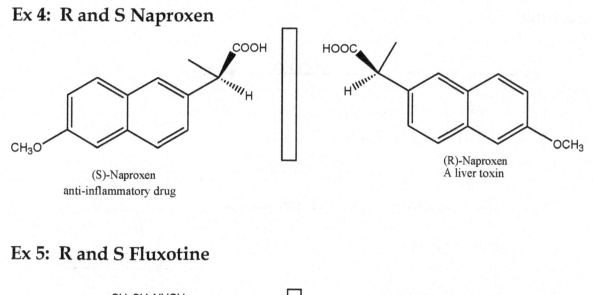

(S)-Naproxen
anti-inflammatory drug

(R)-Naproxen
A liver toxin

Ex 5: R and S Fluxotine

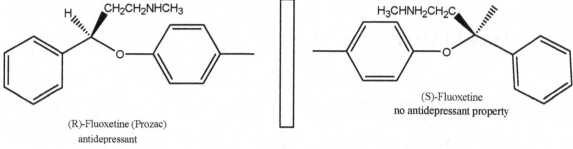

(R)-Fluoxetine (Prozac)
antidepressant

(S)-Fluoxetine
no antidepressant property

- **Note: Resolution is the separation of 2 enantiomers. It is difficult to do. Why? Here are the general steps involved in a racemic resolution:**

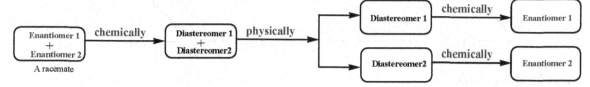

J. PROCHIRALITY

1. DEFINITION

- A **prochiral** substance is an **achiral** molecule that can be converted to **a chiral compound in one step.**

167

Ex: Ethanol

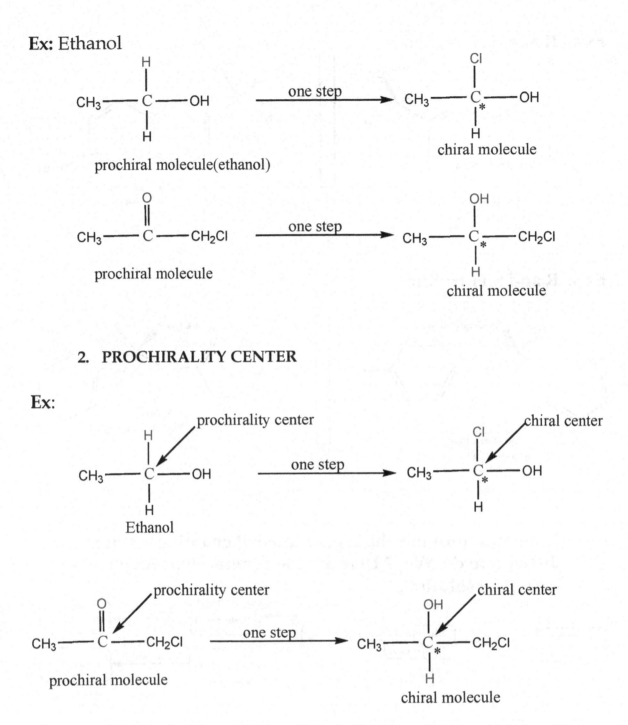

prochiral molecule(ethanol)

one step

chiral molecule

prochiral molecule

one step

chiral molecule

2. PROCHIRALITY CENTER

Ex:

prochirality center

Ethanol

one step

chiral center

prochirality center

one step

chiral center

prochiral molecule

chiral molecule

3. pro-R and pro-S ATOMS ON AN sp³ PROCHIRALITY CENTER:

- If a **prochirality center** is sp³ hybridized, then the two replaceable atoms that lead to chirality can be said to be either *pro-R* or *pro-S*.
- The atom whose replacement gives an R chiral center is *pro-R*.
- The other atom is *pro-S*.

- In the following example, we assume that the replacing atom has a higher **priority** than H.

Ex:

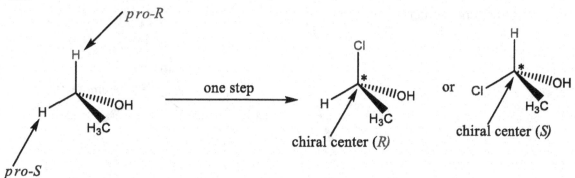

4. si and re FACES ON AN sp² PROCHIRALITY CENTER

- If a prochirality center is sp² hybridized, then we use *re (R)* and *si (S)* to describe the **face** of the trigonal planar sp² carbon on which the reaction occurs.
- Using priority rules, the face that gives a **clockwise** move is called the *re face*.
- The other face is called the *si face*.

Ex:

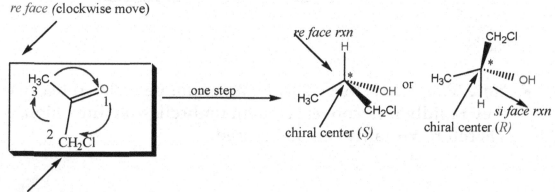

K. CHIRALITY OF OTHER ATOMS

- Chirality is not limited only to tetrahedral sp³ carbon atoms with **4 different groups.** Chiral tetrahedral Si, N, P, and S molecules exist in nature.

Ex:

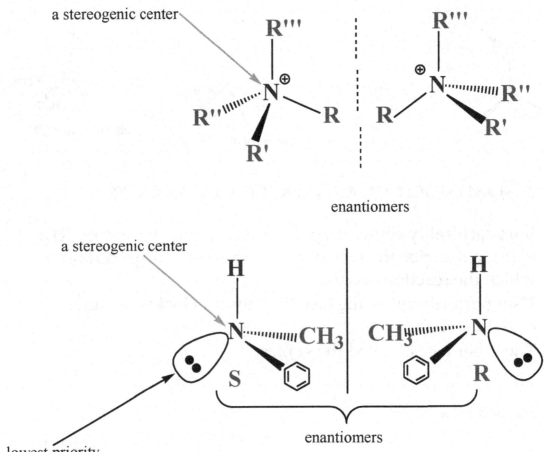

enantiomers

a stereogenic center

lowest priority

enantiomers

- **Note: Practically, these 2 isomers cannot be isolated because they rapidly interconvert thru an umbrella-like inversion. Therefore, chirality can be ignored.**

Ex: Is it R or S?

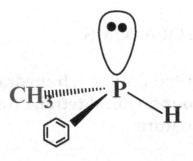

L. ISOMERISM: A SUMMARY

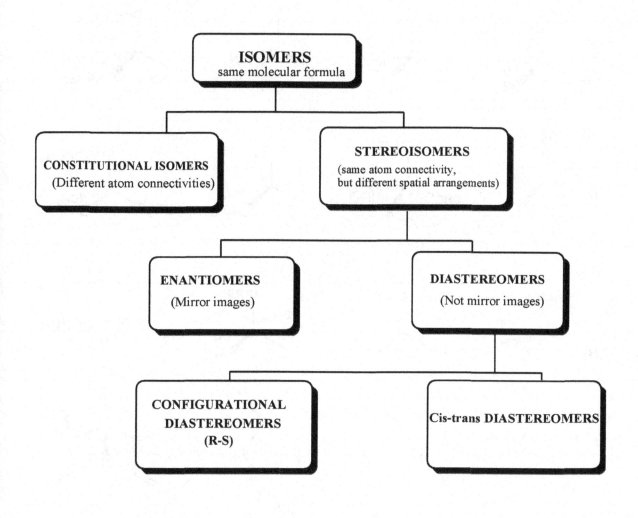

- See Key Concepts, pages _____ - _____ .

• Exercise 1: Assign R-S Configuration to Each Chiral Center

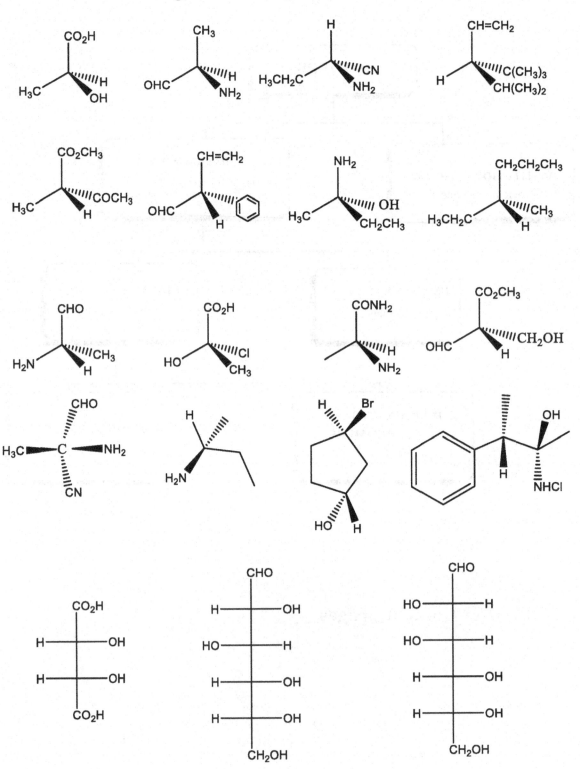

172

- **Exercise 2**: Identify the **stereogenic** centers in cholesterol and calculate the number of possible **stereoisomers** for this compound.

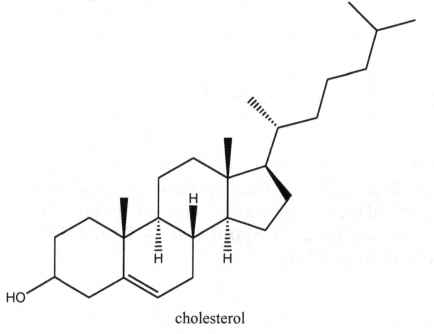

cholesterol

- **Exercise 3**: Assign **D** and **L** configurations to the following 4-carbon molecules: Which compound are enantiomers? Which are diastereomers? Assign R/S configuration to each **stereogenic** carbon.

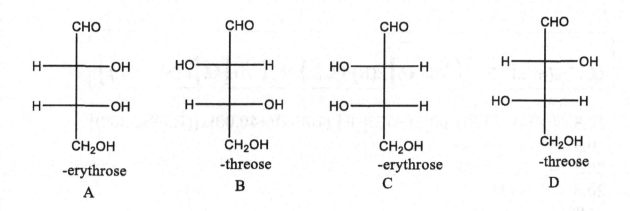

	CHO		CHO		CHO		CHO
H—	—OH	HO—	—H	HO—	—H	H—	—OH
H—	—OH	H—	—OH	HO—	—H	HO—	—H
	CH$_2$OH		CH$_2$OH		CH$_2$OH		CH$_2$OH
	-erythrose		-threose		-erythrose		-threose
	A		B		C		D

- **Exercise 4**: Limonene is a compound found in lemons and oranges. The R is found in oranges and the S in lemons. Assign R and S configurations to both forms.

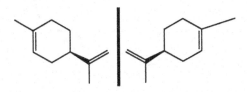

- **Exercise 5:** The observed rotation of a mixture of 2 enantiomers was found to be 20.5°. Calculate the respective % of R and S if l = 10.0 cm and $[\alpha]^D$ of the S enantiomer is +40.0° mL.g⁻¹dm⁻¹. **The respective concentrations are .575 g/mL (for the R) and .200 g/mL (for the S) in the mixture.**

- **Answer:**

$[\alpha]^D$ (R) = -40.0° mL.g⁻¹dm⁻¹
l = 10.0 cm = 1.00 dm

$$\alpha \text{ obs(mixt)} = \left[\left(\% \alpha_{obs}(R)\right) + \left(\% \alpha_{obs}(S)\right)\right]$$

or

$$\alpha \text{ obs(mixt)} = \left[\left(\% [\alpha]_{D(R)} \, l \, C_R\right) + \left(\% [\alpha]_{D(S)} \, l \, C_S\right)\right]$$

20.5= [x (-40.0)x((1.00)x(.575)] + [(100-x)(+40.00)x((1.00)x(.200)]
20.5 = -23 x + 800 -8 x
20.5 = - 31 x + 800
20.5 -800 = -31 x
-7.80 x 10² = -31 x
x = -7.80 x 10²/-31
x = 25.2% R
%S = 100 – 25.2 = 74.8%S

OCHEM UNIT 8: AN OVERVIEW OF ORGANIC CHEMICAL REACTIONS

A. WRITING CHEMICAL EQUATIONS IN OCHEM

1. ONE-STEP REACTIONS

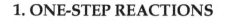

Ex:

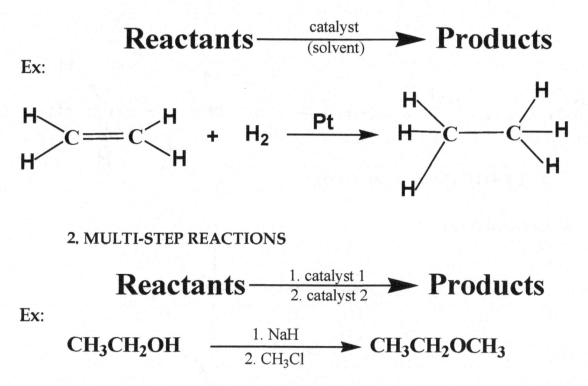

2. MULTI-STEP REACTIONS

Ex:

$$CH_3CH_2OH \xrightarrow[\text{2. } CH_3Cl]{\text{1. NaH}} CH_3CH_2OCH_3$$

B. TYPES OF CHEMICAL REACTIONS IN OCHEM

1. INTRODUCTION

- There are 4 major types of organic reactions:
    - Addition reactions
    - Elimination reactions
    - Substitution reactions
    - Rearrangement reactions

- Read pages _____ – _____.

175

2. ADDITION REACTIONS

- General reaction:

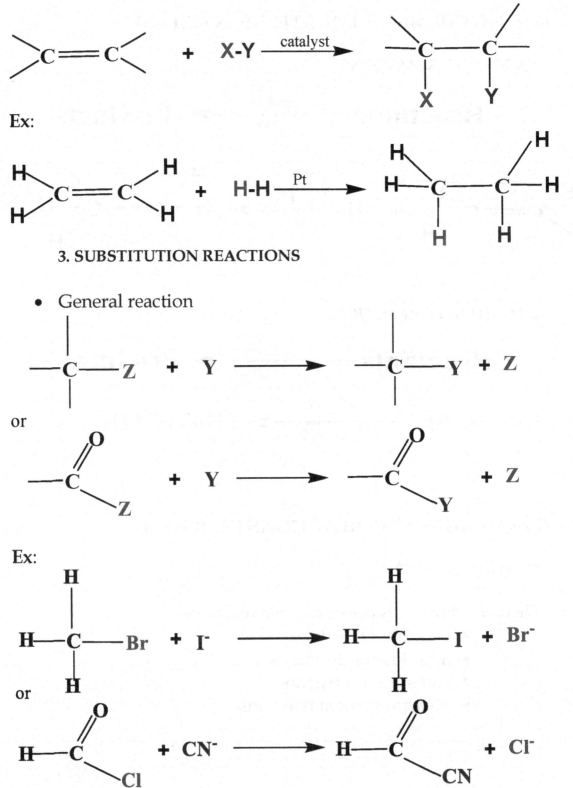

Ex:

3. SUBSTITUTION REACTIONS

- General reaction

or

Ex:

or

176

4. ELIMINATION REACTIONS

- ## General Reaction

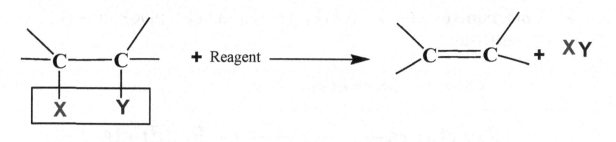

Ex:

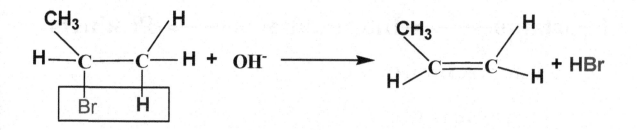

5. REARRANGEMENT REACTIONS: INTERNAL REORGANIZATION

Ex: Tautomerization

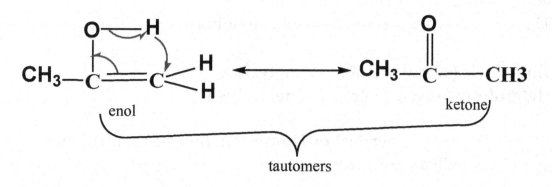

- **Do Problem _____, page _____.**

C. MECHANISMS OF CHEMICAL REACTIONS

1. INTRODUCTION

- A **mechanism** tells about **how** a reaction takes place: one-step or multi-step.

a. One-Step Mechanism

$$\textbf{Reactants} \longrightarrow \textbf{Products}$$

b. Multi-Step Mechanism

$$\textbf{Reactants} \longrightarrow \textbf{Intermediates} \longrightarrow \textbf{Products}$$

2. REACTION MECHANISM

a. Introduction

$$\textbf{Reactants} \longrightarrow \textbf{Products}$$

b. Bond-Breaking (Cleavage)

- Bond-breaking is an **endothermic** process.
- There are **2 ways** of breaking a covalent bond:

- **Homolytic** (symmetrical) = **homolysis**
- **Heterolytic** (unsymmetrical) = **heterolysis**

i. Symmetrical or Homolytic Bond-Breaking (or Cleavage): Homolysis

- In this kind of bond-breaking, **2 free radicals are produced.**

- **General Reaction:**

Ex:

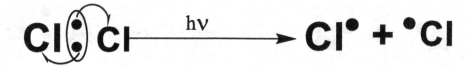

- **Note: Radicals are very reactive.**

    ii. Unsymmetrical or Heterolytic Bond Breaking (or Cleavage): Heterolysis

- In this kind of bond-breaking, **2 ions are produced.**

- **General Reaction:**

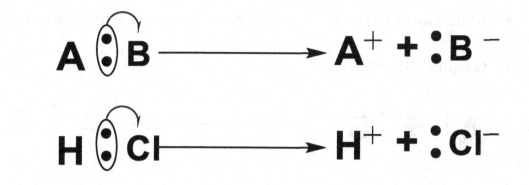

Ex:

    c. Bond-Forming

- Bond-forming is an **exothermic** process.
- There are **2 ways** of forming a covalent bond:

- **Homogenic** (symmetrical)
- **Heterogenic** (unsymmetrical)

i. Symmetrical or Homogenic Bond-Forming

- In this kind of bond-forming, **2 free radicals combine to give products.**

- **General Reaction:**

Ex:

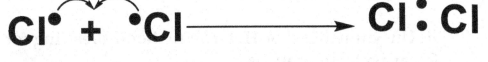

- **Note: a reaction that involves free radicals is called a <u>radical reaction.</u>**

ii. Unsymmetrical or Heterogenic Bond-Forming

- In this kind of bond-forming, **2 ions combine to give products.**

- **General Reaction:**

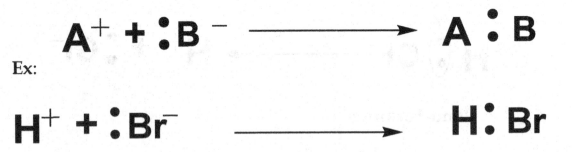

- **Note: a reaction that involves ions is called a <u>polar reaction</u>.**

- **Read pages _____ – _____. Do all problems.**

3. USING ARROWS IN RADICAL AND POLAR REACTIONS

- See Table _____, page_____.
- Become familiar with the symbols used.
- Do problems on page _____.

D. REACTION INTERMEDIATES

1. INTRODUCTION

- There are **4 types** of reaction intermediates:
  - Free radicals
  - Carbocations
  - Carbanions
  - Carbenes

2. FREE RADICALS

a. General Structure

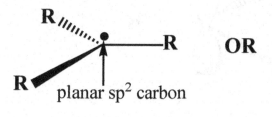

planar sp^2 carbon

OR

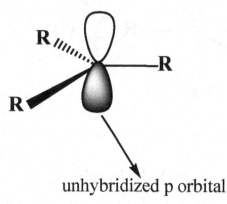

unhybridized p orbital

b. VB Sketch

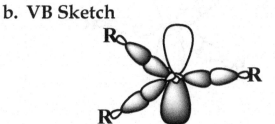

c. Stability

- Resonance stabilized
- Overlap with the p orbital of a π bond **(hyperconjugation)**

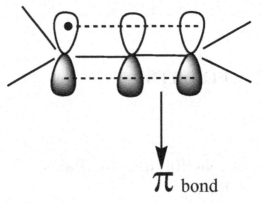

π bond

d. Reactivity

- A **free radical** is electron deficient and therefore is **an electrophile (electron-seeking)**.

3. CARBOCATIONS

a. General Structure

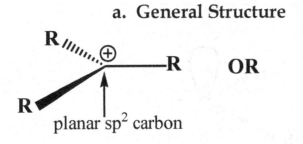

planar sp^2 carbon

OR

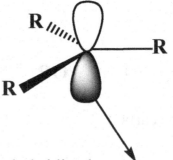

unhybridized vacant p orbital

b. VB Sketch

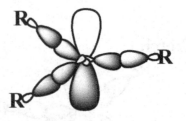

c. Stability

- Inductive effect
- Hyperconjugation

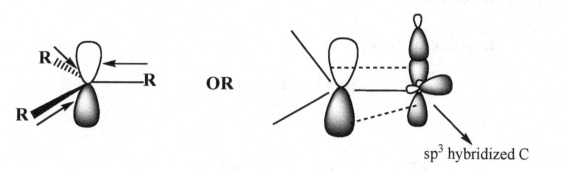

OR

sp^3 hybridized C

d. Reactivity

- A **carbocation** is positive and **electron deficient**, and thus is an **electrophile (electron-seeking)**.

4. CARBANIONS

a. General Structure

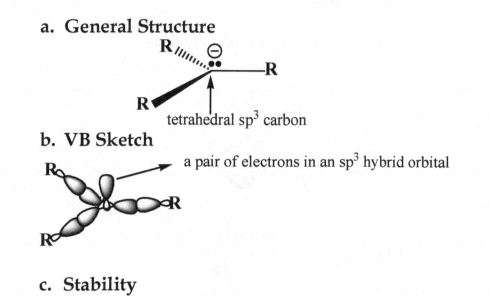

tetrahedral sp^3 carbon

b. VB Sketch

a pair of electrons in an sp^3 hybrid orbital

c. Stability

- Fairly stable, but destabilized by **electron-donating groups (EDG)**

Ex: CH_3-

d. Reactivity

- A **carbanion** is negative and therefore is **a good nucleophile and a good Lewis base (electron pair donor).**

5. CARBENES

a. General Structure

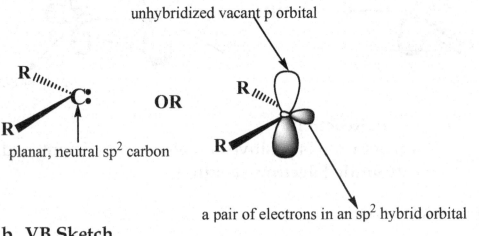

OR

planar, neutral sp² carbon

unhybridized vacant p orbital

a pair of electrons in an sp² hybrid orbital

b. VB Sketch

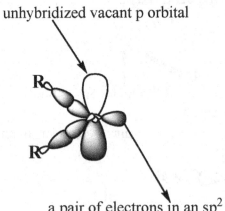

unhybridized vacant p orbital

a pair of electrons in an sp² hybrid orbital

c. Stability

- Carbenes are **divalent** carbon compounds (**acetylenic** carbons) that are very unstable.

d. Reactivity

- No octet→ react as **electrophiles (electron-seeking)**; a Lewis acid = an **electron pair acceptor**.

E. BOND DISSOCIATION ENERGIES

1. DEFINITION

- The bond dissociation energy (BDE) of a covalent bond is the energy required to break a covalent bond in the gaseous state (**homolytic bond-breaking**).

$$A : B \longrightarrow A\cdot + B\cdot$$

- See Table _____, page _____.

- **Note: The higher the BDE, the stronger the bond.**

Bond	C-H	Cl-Cl	C-Cl	H-Cl	H-I	C-I	C-Br	H-F	Br-Br	C-F
BDE	435	242	351	431	297	234	293	569	192	456

2. TRENDS IN BDE IN C-X

$$CH_3\text{-}F > CH_3\text{-}Cl > CH_3\text{-}Br > CH_3\text{-}I$$

3. CALCULATION USING BDE

Reactants ⟶ Products

$$\Delta H^\circ_{rxn} = \sum D(\text{broken bonds}) - \sum D(\text{bonds formed})$$

Ex: $CH_4 + Cl_2 \longrightarrow CH_3Cl + HCl$

- Do Problems _____ page _____ and _____ page _____ .

- Do problems on pages _____ - _____ .

F. THERMODYNAMICS IN OCHEM

1. INTRODUCTION: thermodynamics vs. kinetics

- Thermodynamics = energies of reactants and products
  = who is favored at equilibrium?
- Kinetics = reaction rates
  = how fast products are being formed.

2. THE EQUILIBRIUM CONSTANT

Given the following reaction at equilibrium

$$aA + bB \rightleftharpoons cC + dD$$

$$K_{eq} = \frac{[C]^c[D]^d}{[A]^a[B]^b}$$

- **Question: Who is favored at equilibrium?**

- If $K_{eq} > 1$, then products are favored at equilibrium. In other words, the forward reaction is favored at equilibrium.
- If $K_{eq} < 1$, then reactants are favored at equilibrium. In other words, the reverse reaction is favored at equilibrium.
- If $K_{eq} = 1$, then reactants and products are equally favored at equilibrium.

3. GIBBS FREE ENERGY CHANGE (ΔG°)

- ΔG° = Standard **free energy change** that occurs during a chemical reaction.

- If $\Delta G^\circ < 0$, have a **spontaneous (exergonic) reaction.**

- If $\Delta G^\circ > 0$, have a **nonspontaneous (endergonic) reaction.**
- **Note:** $\Delta G^\circ = G^\circ$ **(products) -** G° **(reactants). So** $\Delta G^\circ < 0$ **means that energy of products is lower than energy of reactants.**
- **So** $\Delta G^\circ > 0$ **means that energy of products is higher than energy of reactants.**

4. RELATIONSHIP BETWEEN ΔG° AND K_{eq}.

$$\boxed{\mathbf{\Delta G^\circ = -RT \ln K_{eq}}}$$

- **Recall $\ln x$:**

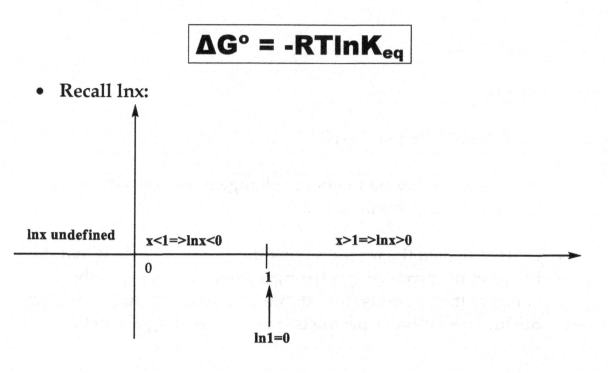

- If $Keq > 1$, then $\ln Keq > 0$ ➜ $\Delta G^\circ < 0$ ➜ **energy is released during the reaction.** The reaction is **exergonic.**

- If $Keq < 1$, then $\ln Keq < 0$ ➜ $\Delta G^\circ > 0$ ➜ **energy is absorbed during the reaction.** The reaction is **endergonic.**

- See Fig. _____, page _____.

- Read pages _____ - _____.

- See Table _____, page _____.

- Do problems on page _____.

- **Note:** Given $aA \rightleftharpoons bB$

- Keq = y means at equilibrium, there is y times more of B than A

  Ex: Keq = 1×10^2 means at equilibrium, there is 100 times more of B than A.

- Read pages _____ and _____.

- Do problem on page _____.

5. ENTHALPY CHANGE: $\Delta H°$

- $\Delta H°$ measures **the bond energy change** that occurs during a reaction at **constant pressure**.

- If $\Delta H° < 0$, then the reaction is **exothermic**. This means that the bonds of the products are **stronger (more stable)** than the bonds of the reactants (**less stable**). Therefore, **heat is released during formation of products. See Energy diagram below:**

- If $\Delta H < 0 \rightarrow H_{products} < H_{reactants}$

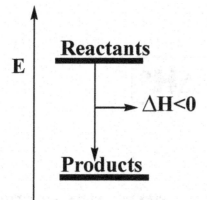

- If $\Delta H° > 0$, then the reaction is **endothermic**. This means that the bonds of the products are **weaker (less stable)** than the bonds of the reactants (**more stable**). In this case, **heat is absorbed before formation of products. See energy diagram below:**

- $\Delta H > 0 \rightarrow H_{products} > H_{reactants}$

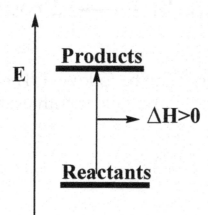

6. ENTROPY CHANGE: $\Delta S°$

- $\Delta S°$ measures the **energy dispersal** that occurs during a reaction.

7. RELATING $\Delta G°$, $\Delta H°$, $\Delta S°$

$$\Delta G° = \Delta H° - T\Delta S°$$

- At low temperatures, $-T\Delta S^o$ is negligible. The relationship becomes

$$\boxed{\Delta G^o = \Delta H^o}$$

- Read page _____.

8. ENERGY DIAGRAMS AND TRANSITION STATES

a. Introduction

- Suppose, we have a reaction such as

$$A + B \longrightarrow [A\ldots\ldots B] \longrightarrow Products$$

- As the reaction proceeds, it can be followed. The result is called an **energy diagram**. Let's take a look at both **exergonic and endergonic reactions.**

b. Exergonic Reactions: $\Delta G° < 0$

- The free energy of the products is **lower** than the free energy of the reactants.

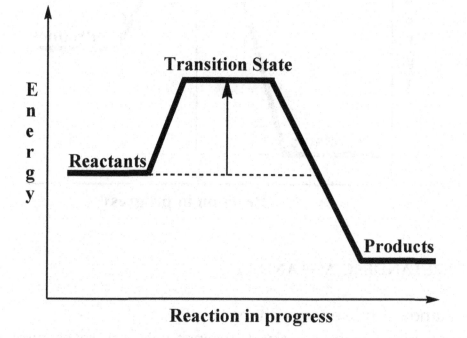

c. Endergonic Reactions: $\Delta G° > 0$

- The free energy of the products is **higher** than the free energy of the reactants.

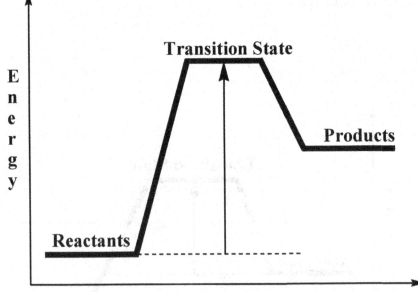

9. UNDERSTANDING $\Delta G°$ AND E_a

- $\Delta G°$ = standard free energy change.
- = accounts for the energy difference between reactants and products.
- Review Section 3.

- E_a = activation energy
- = energy required to go from reactants to the transition state.
- = tells about how fast a reaction is occurring: a **high activation energy** means a **slower reaction**; a **low activation energy** means a **faster reaction**.

- See Fig. _____, page _____.
- See Fig. _____, page _____.
- Do problems on pages _____, _____, and _____.

10. ENERGY DIAGRAMS FOR 2-STEP REACTION MECHANISMS

- Suppose we have a **2-step** reaction:
- 1st step endergonic
- 2nd step exergonic

$$A + B \longrightarrow C \longrightarrow D$$

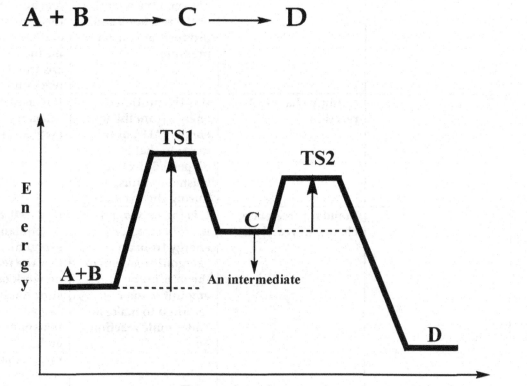

- See Fig. _____, page _____.

- Read pages _____ – _____.

- Do all Problems on page _____.

11. THERMODYNAMIC PARAMETERS: A SUMMARY

Symbol	Name	Meaning	Use
$\Delta H°$	Standard heat (or enthalpy) change of reaction	Total energy change that occurs during a reaction; it is the energy difference between the energies of the reactants and products at constant pressure	If $\Delta H° < 0$, the reaction is exothermic; this means heat is given off during the reaction. If $\Delta H° > 0$, the reaction is endothermic; this means heat is absorbed during the reaction
$\Delta S°$	Entropy change of reaction	It is the portion of energy from the total energy ($\Delta H°$) of the reaction that is dispersed (lost or wasted or unused) during the reaction	It is usually used to calculate $\Delta G°$. ($\Delta G° = \Delta H° - T\Delta S°$)
$\Delta G°$	Standard free energy change of reaction	It is the maximum useful amount of energy from an exergonic reaction or the minimum amount of energy required to make an endergonic reaction go	If $\Delta G° < 0$, the reaction is spontaneous and exergonic; this means that the reaction can do work on its surroundings. If $\Delta G° > 0$, the reaction is nonspontaneous and endergonic; this means that the reaction cannot do work on its surroundings.
Keq	Equilibrium constant	Tells about who (reactants and products) is favored at equilibrium	If Keq>1 => products are favored at equilibrium; If Keq<1 => reactants are favored at equilibrium; If Keq=1 => reactants and products are equally favored at equilibrium.

G. CHEMICAL KINETICS

1. DEFINITION

- **Chemical kinetics** is the field of Chemistry that studies the **rates** of chemical reactions and their **mechanisms**.
- **Recall: low activation energy, faster reaction; high activation energy, slower reaction.**

2. FACTORS AFFECTING REACTION RATES

- Concentration.
- Temperature.
- Energy of activation.

- **Note: The rate of a reaction does not depend on thermodynamic parameters ΔG^o, ΔH^o, ΔS^o or K_{eq}. However, it depends on Ea.**

3. THE RATE LAW

a. Definition

- A **rate law** is an **equation or a relationship** that expresses how the rate of a reaction depends on the concentrations of the reactants in a chemical reaction.

- **Note: For a given reaction, the rate law is always the same regardless of the concentrations of the reactants.**

b. Expression of the Rate Law

- For the reaction: $A + B \longrightarrow$ **Products**

The rate law of this reaction is of the form:

$$\boxed{Rate = k[A]^m[B]^n}$$

- m = order of the reaction in A
- n = order of the reaction in B
- m + n = overall order of the reaction.

- If m = 1 ➜ 1st – order in [A]
- If n = 1 ➜ 1st – order in [B]
  Ex: Rate = k[A][B]   overall = 1+1 = 2
- k = rate constant
- **Note: The higher the k, the faster the reaction, and vice versa.**

4. MULTI-STEP REACTIONS AND RATE DETERMINING STEP

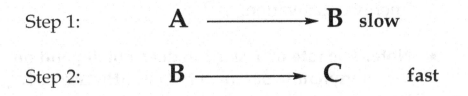

Step 1: $A \longrightarrow B$ slow

Step 2: $B \longrightarrow C$ fast

- **Slowest** step is **rate determining step.**

Rate = $k_1[A]$

- Read pages _____ – _____.

- Do problems on pages _____, _____, and _____.

5. CATALYSTS

a. Definition

- A catalyst is a substance that **speeds up** a reaction without undergoing a **net chemical change itself**.

Ex: hydrogenation of ethene

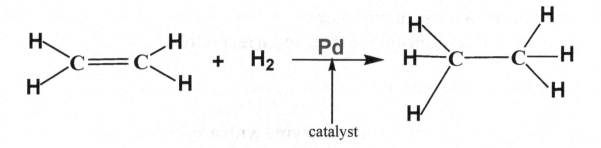

catalyst

b. Catalysis and the Activation Energy

- A **catalyst** lowers the overall activation energy of a reaction or provides **an alternative** pathway with a **lower activation energy.**

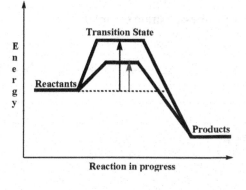

- See Fig. _____, page _____.

- **Note: Although a catalyst can help a reaction reach equilibrium faster, it does not affect the position of an equilibrium. Indeed, it does not affect the concentrations of reactants and products at equilibrium.**

- Do problem on page _____.

6. ENZYMES AND INHIBITORS: ENZYMOLOGY

a. Introduction

- **Enzymes** are biological catalysts made of **protein molecules.** An **inhibitor** deactivates an enzyme's catalytic activity by preventing it from forming a "complex" (ES) intermediate with the **substrate** (reactant).

- There are **3 types** of inhibitors:
- **competitive, noncompetitive, and irreversible.**

b. Enzymatic Activity

- The catalytic reaction of an enzyme with a substrate can be represented by the following **2-step reaction**.

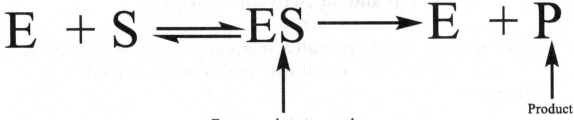

Enzyme-substrate complex

Product

- Read pages _____ - _____. See Lactase in action on page _____.

- See Fig. _____, page _____.

Ex: Catalase

$$2H_2O_2 \xrightarrow{\text{catalase}} 2H_2O + O_2$$

- Read pages _____ - _____.

c. Enzymatic Models: 2

  i: The Lock-and-Key Model

- In this model, substrate and enzyme have rigid, inflexible **complementary** shapes like a door key and its lock.

ii: The Induced-Fit Model

- Unlike the lock-and-key model, this model stipulates that substrate and enzyme have flexible, **similar** shapes that undergo **conformational change** upon contact to adjust to each other.

d. Inhibitor Action

- An **inhibitor** actually prevents the formation of the **ES complex** by forming an enzyme-inhibitor complex (EI) as follows:

$$E + S + I \rightleftharpoons EI + S$$

inhibitor

unreactive substrate

Enzyme-Inhibitor complex

- **A competitive inhibitor** has a structure **similar** to that of the substrate. As a result, it **lodges** itself to the reacting site of the enzyme thus blocking it. I competes with the substrate for the enzyme when both are present.
  Ex: Malonate cripples the enzymatic activity of **succinate dehydrogenase** by binding to its reacting site.
- **A noncompetive inhibitor** does not tie up the binding site of the enzyme as it attaches itself at a site other than the binding site. However, the EI distorts the shape of the enzyme preventing it from doing its job.

Ex: Heavy metals (Hg^{2+}, Pb^{2+})

- **An irreversible inhibitor** causes an enzyme to lose its catalytic properties by reacting with an amino acid side chain. In other words, it destroys the enzyme ability to catalyze a reaction.
  Ex: Toxic substances, poisons, venoms

- **See Key Concepts, pages _____ - _____.**

OCHEM I UNIT 9: ALKYL HALIDES

A. INTRODUCTION

1. GENERAL STRUCTURE

R — X   or

- R = alkyl group

- X = F, Cl, Br, I

Ex: CH_3Cl

2. KINDS OF HALIDES

a. Simple Alkyl halides

Ex: CH_3CH_2Br

b. Vinyl Halides

Ex: $CH_2=CH-Br$

c. Aryl Halides

Ex:

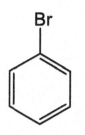

d. Allylic Halides

Ex: $CH_2=CHCH_2-Br$

e. Benzylic Halides

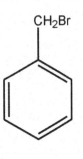

Ex:

3. CLASSIFICATION OF ALKYL HALIDES: 4 Kinds

- Read pages _____ - _____.

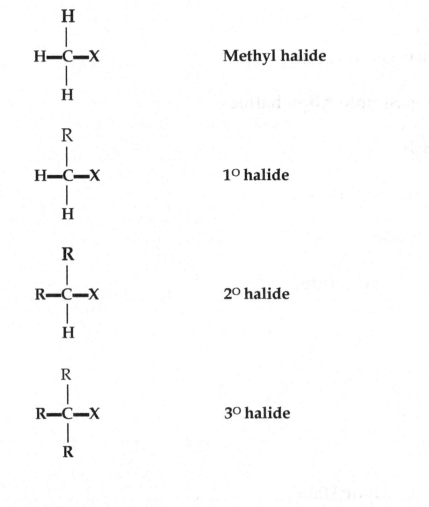

- Do problem on page _____.

B. NOMENCLATURE OF ALKYL HALIDES

1. IUPAC

- Naming is the same as in the case of the alkanes, except that **halo prefixes** are used. See **table in Unit 5.**

$$-ine \longrightarrow -o$$

- These are the halo prefixes:

element	group
fluor*ine*	fluor*o*
chlor*ine*	chlor*o*
brom*ine*	brom*o*
iod*ine*	iod*o*

Ex: Name the following compounds:

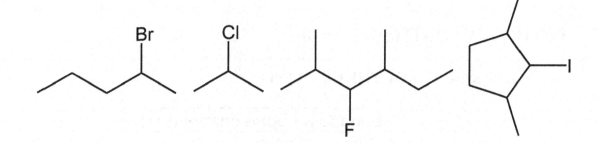

Ex: Write a structure for 2,5-dichloro-3-iodoheptane

- Please, see page _____.

- Do Problems on page _____.

2. COMMON NAMES

- Common names are usually used for simple alkyl halides. They are named as **alkyl halides**.

> **Name of alkyl group + name of anion of the halogen**

Ex:

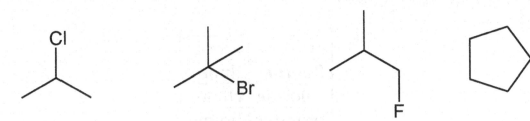

Ex: Write a structure for isopropyl bromide.

- Do Problems _____, _____, page _____.

C. THE C-X BOND (See Unit 8)

1. STRENGTH OF THE C-X BOND

Bond C-X	Bond energy (kJ)
C-F	456
C-Cl	351
C-Br	293
C-I	234

- Note: The increasing order of strength is: C-I < C-Br < C-Cl < C-F.
- C – F = strongest
- C – I = weakest

2. THE C – X BOND POLARITY

- The C – X bond is a **polar** bond.

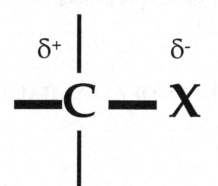

- See electrostatic potential plots, page _____.

D. SYNTHESIS OF ALKYL HALIDES

1. HALOGENATION OF THE ALKANES

- **Review Unit 5.**

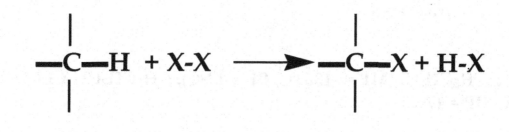

Ex:

$$CH_3CH_3 + Br_2 \longrightarrow CH_3CH_2Br + HBr$$

E. REACTIONS OF ALKYL HALIDES

- **Note: The Chemistry of alkyl halides is dominated by nucleophilic substitution reactions.**
- **Will see in Unit 10.**

F. PHYSICAL PROPERTIES OF ALKYL HALIDES

1. INTRODUCTION

- Alkyl halides are **polar substances**. Therefore, they can form **dipole-dipole interactions** between their molecules. Since they **do not contain OH groups**, they **cannot form H bonds**.

2. MELTING AND BOILING POINTS OF ALKYL HALIDES

- Alkyl halides have higher BP and MP than comparable nonpolar alkanes.

Ex: CH_3CH_3 (BP = - 89°C) and CH_3CH_2Br (BP = 39°C)

- **Note: MP and BP increase with increasing size of R: larger surface area.**

Ex: CH_3CH_2Cl (MP = -136 °C; BP = 12°C); $CH_3CH_2CH_2Cl$ (MP = -123 °C; BP = 47°C)

- **Note: MP and BP increase with increasing size of X due to polarizability.**

Ex: CH_3CH_2Cl (MP = -136 °C; BP = 12°C); CH_3CH_2Br (MP = -119 °C; BP = 39°C); Br is more polarizable than Cl.

3. SOLUBILITY

- R-X is **soluble in organic solvents.**
- R-X is **insoluble in water.**

- Do problem on page.

- See Table on page _____

G. IMPORTANT ALKYL HALIDES

1. SOLVENTS

- Simple alkyl halides are liquid at room temperature and are used as solvents although they are toxic.

Ex: CH_3Cl (chloroform); CH_2Cl_2 (methylene chloride)

2. CFCs and HFCs

a. Definition
- CFCs = chlorofluorocarbons = freons
- Used as refrigerants and aerosol propellants
- **$CFCl_3$ = Freon 11**
- **CF_2Cl_2 = Freon 12**
- **CF_3Cl = Freon 13**
- Read about DDT = DichloroDiphenylTrichloroethane a pesticide on page _____

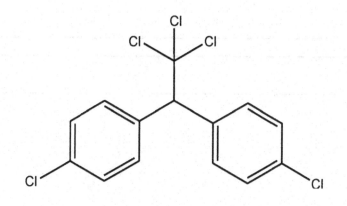

207

b. How Do CFCs Deplete the Ozone Layer?

- **CFCs** are believed to be the culprits in the depletion of Earth's atmosphere's **ozone layer**.
- **See diagram below:**

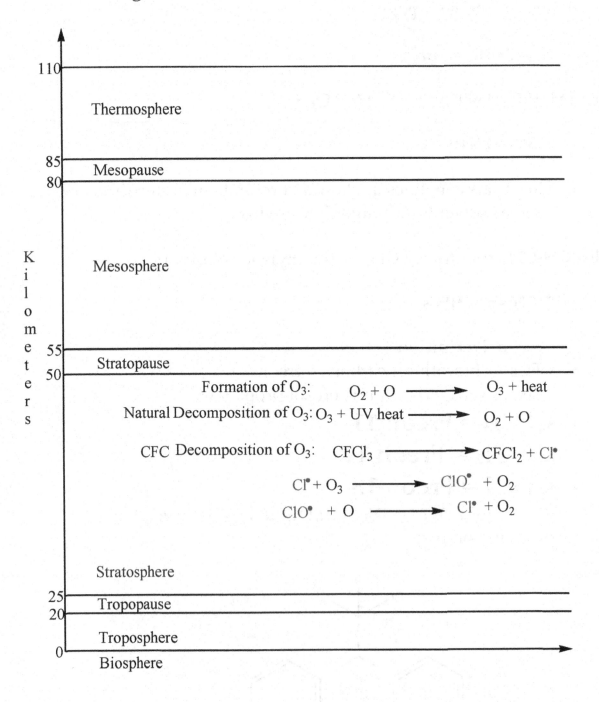

- HCFCs and HFCs = **hydrochlorofluorocarbons** and **hydrofluorocarbons** = substitutes of CFCs
- Used as refrigerants and aerosol propellants
- **CFH_3**
- **CF_2H_2**
- **CF_3H**

- **Note: HCFCs and HFCs are decomposed in the troposphere before they reach the stratosphere.**

- See example below.

Ex:

$$HO\cdot + HCHFCF_3 \longrightarrow H_2O + \cdot CHFCF_3$$

3. ANESTHETICS

- Halothane is used in surgery as **anesthetic: $CF_3CHClBr$**

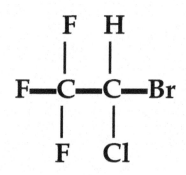

4. POLYMERS

- Teflon
- PVC

- Read pages _____ - _____.

- See Key Concepts, pages _____ - _____.

OCHEM UNIT 10: NUCLEOPHILIC SUBSTITUTION REACTIONS

A. INTRODUCTION

1. ALKYL HALIDES REVISITED

R — X   where

- R = alkyl group

- X = F, Cl, Br, I, and others.
- Electrophile = electron-poor
    = electron-loving
- Nucleophile = positive loving
    = electron rich
- **Note:** An **electrophile** reacts with a **nucleophile** in a polar reaction.
- R-X = the substrate
- X = the leaving group

2. TYPES OF POLAR REACTIONS

a. Reactions of Nucleophiles

- Nucleophiles undergo **two** types of **polar reactions**:
- **Substitution reactions**
- **Elimination reactions.**

b. Substitution Reactions

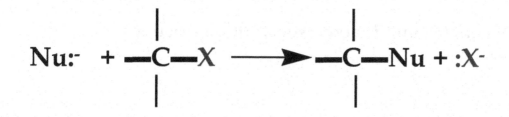

Ex:

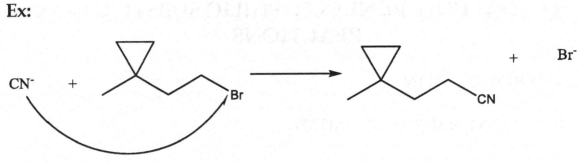

a substitution reaction

c. Elimination Reactions: Need a Base

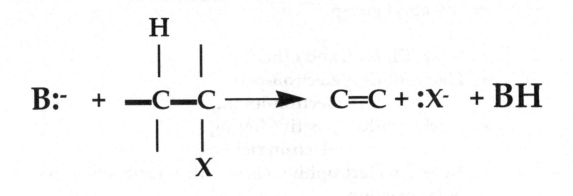

Ex:

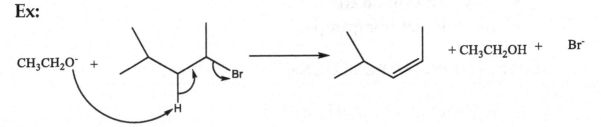

an elimination reaction

- Read pages _____ - _____.

- Do problems on page _____.

- Note: This unit is about substitution reactions.

212

3. TYPES OF SUBSTITUTION REACTIONS

- There are **two types** of **substitution** reactions:

  - SN1.
  - SN2.

B. SN2 REACTIONS

1. INTRODUCTION

$$Nu:^- \ + \ R\!-\!X \longrightarrow R\!-\!Nu + :X^-$$

Ex:

$$OH:^- + CH_3\!-\!Br \longrightarrow CH_3\!-\!OH + :Br^-$$

2. THE KINETICS OF SN2

- Recall:

$$Nu:^- \ + \ R\!-\!X \longrightarrow R\!-\!Nu + :X^-$$

$$\boxed{Rate = k[nucleophile][substrate]}$$

$$\boxed{Rate = k[Nu][R\text{-}X]}$$

- Note: The reaction is bimolecular (a second-order reaction). In other words, the rate of the reaction depends on both nucleophile and substrate.
- # S = substitution
- # N = Nucleophilic
- # 2 = bimolecular

Ex:

$$\text{OH:}^- + \text{CH}_3\text{—I} \longrightarrow \text{CH}_3\text{—OH} + \text{:I}^-$$

$$\text{Rate} = k[\text{CH}_3\text{I}][\text{OH}^-]$$

- Note: If we double [CH₃I], the rate doubles. If we double [OH⁻], the rate also doubles.

3. REACTION MECHANISM

- The SN2 reaction occurs in **one single step**. We have **concerted bond-breaking and bond-forming**. In other words, bond-breaking and bond-forming occur at the **same time**. We have a **nucleophilic backside attack**.

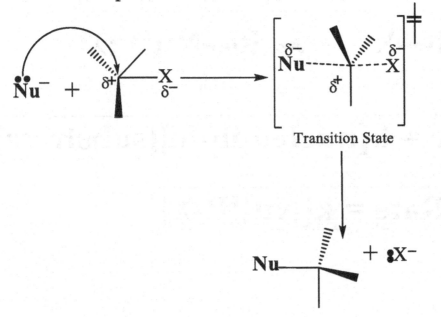

Transition State

214

Ex:

$$CN\!:^- + CH_3\!-\!Br \longrightarrow CH_3CN + Br^-$$

- Mechanism:

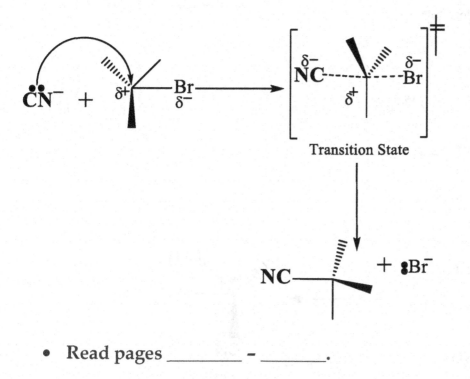

Transition State

- Read pages _____ – _____.

- Do problems on page _____.

4. SN2 REACTIONS INVOLVING CHIRAL SUBSTRATES

- SN2 occurs with **inversion of configuration (Walden Inversion)** when **chiral substrates** react.
- **Why?** Because of **backside attack** by the nucleophile.

- **R-substrate ⟶ S-product**
- **S-substrate ⟶ R-product**
- **cis-substrate ⟶ trans -product**
- **trans-substrate ⟶ cis-product**

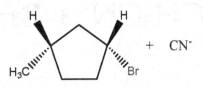 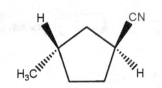

cis-1-Bromo-3-methylcyclopentane

trans-1-Cyano-3-methylcyclopentane

- See Fig. _____, page _____.

- Read pages _____ - _____.

Ex:

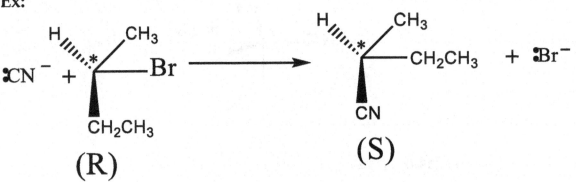

(R)   (S)

C. CHARACTERISTICS OF SN2 REACTIONS

1. INTRODUCTION

- The **rate** of SN2 depends on **several factors:**
- **The size of the substrate.**
- **The strength of the nucleophile.**
- **The leaving group.**
- **The solvent.**

- Recall: E_a = activation energy

- Small E_a = fast exergonic reaction. Higher reactant energy
- Small E_a = faster reaction.

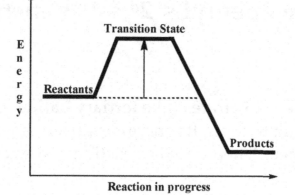

- Large E_a = slow endergonic reaction.
- Higher transition state energy
- Larger E_a = slower reaction.

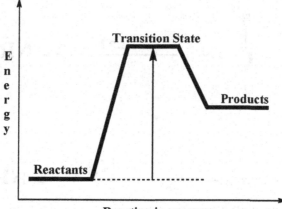

- See Fig. _____, page _____.

2. EFFECT OF THE SUBSTRATE (RX) ON THE RATE OF SN2

- The increasing order of the rate of **SN2** is:

$3°$ < neopentyl < $2°$ < $1°$ < methyl halide

Why?

- For **bulky substrates** (like **tertiary halides**), the transition state is difficult to form. Its energy is **raised** due to crowding (**steric hindrance**). Therefore the reaction is **slower**.

- See Fig. _____, page _____.

- **Note: $3°$, vinylic, and aryl substrates are unreactive toward SN2 because of steric hindrance.**

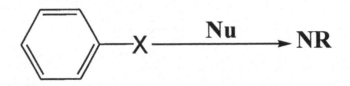

$$H_2C{=}CH{-}X \xrightarrow{\text{Nu}} NR$$

vinyl halide

$$\xrightarrow{\text{Nu}} NR$$

aryl halide

3. EFFECT OF THE NUCLEOPHILE ON THE RATE OF SN2

a. Nucleophiles with the Same Reacting Atom

- The **most basic** nucleophile has the **highest reactivity.**

Ex: Which one is the best nucleophile in SN2, H_2O or OH^-? **OH^-(more basic).**

b. Periodic Trends in Nucleophilicity

- Nucleophilicity **increases** within a **group.**

Ex: $$F^- < Cl^- < Br^- < I^-$$

$$OH^- < SH^- < SeH^-$$

$$(CH_3CH_2)_3N: < (CH_3CH_2)_3P:$$

c. Negative vs. Neutral Nucleophile

- **Negative (more basic)** nucleophiles are more reactive than **neutral** ones.

Ex:

$$OH^- > H_2O$$

$$SH^- > (CH_3)_2S$$

$$NH_2^- > NH_3$$

d. Electronegativity of the Reacting Atom

- Nucleophilicity **decreases** with the **electronegativity** of the reacting atom.

Ex:

$$OH^- > F^-$$

$$NH_3 > H_2O$$

$$(CH_3CH_2)_3P: > (CH_3CH_2)_2S:$$

4. EFFECT OF THE LEAVING GROUP ON THE RATE OF SN2

a. Introduction

- Recall:

$$Nu:^- \ + \ R\!\!-\!\!X \longrightarrow R\!\!-\!\!Nu + :X^-$$

$$\boxed{Rate = k[Nu][R\text{-}X]}$$

- Note: The rate of SN2 depends on the leaving group.

b. Good Leaving Groups

- A good leaving group must be **electron-withdrawing** so that the C can have a **positive partial charge**.
- Note: A good leaving group must contain **O, halogen, S, or N.**

- The conjugate bases of the strongest acids are the best leaving groups. Why? Because these ions are very stable.

- The increasing order of reactivity is:

$$OH^- < NH_2^- < F^- < Cl^- < Br^- < I^- < TosO^-$$

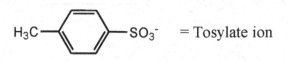 = Tosylate ion

- Finally, a good leaving group must be **polarizable. It must have the ability to bond loosely with C while leaving.**

Ex: I^-

- See Tables _____ and _____, page _____. Read pages _____ - _____.

c. Poor (Bad) Leaving Groups

$$OH^-, NH_2^-, F^-, RO^-$$

- Therefore, substrates R-X = R-F, R-OH, R-OR', R-NH$_2$ do not undergo SN2 reactions.

- $Nu{:}^- + CH_3{-}NH_2 \longrightarrow NR$
- $Nu{:}^- + CH_3{-}OH \longrightarrow NR$
- $Nu{:}^- + CH_3{-}F \longrightarrow NR$
- $Nu{:}^- + CH_3CH_2{-}OCH_3 \longrightarrow NR$

d. Converting a Bad Leaving Group (-OH, -NH₂, HOR) to A Good One

- One can convert bad –OH to good H_2O in acid = a good leaving group as follows:

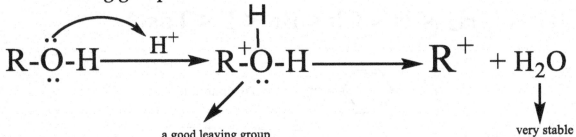

a good leaving group

very stable

Ex:

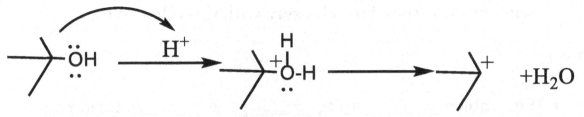

$+H_2O$

- One can also convert bad –OH to good leaving group -OTs using TsCl as follows:

Ex:

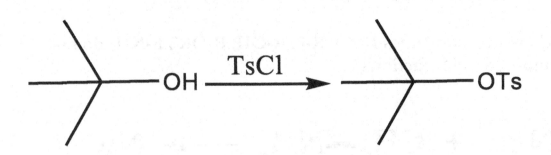

5. EFFECT OF THE SOLVENT ON THE RATE OF SN2

a. Introduction

- The affinity of a nucleophile for C also depends on the **solvent.**

Ex: For the reaction

$$Nu:^- \ + \ R\!\!-\!\!Br \ \longrightarrow \ R\!\!-\!\!Nu + :Br^-$$

- The increasing order of reactivity for a series of nucleophiles in **aqueous ethanol** is:

$H_2O < F^- < CH_3CO_2^- < NH_3 < Cl^- < OH^-, Br^-, N_3^- < CH_3O^- < I^- < CN^- = HS^-$

- **Conclusion: The solvent is very important in SN2 reactions.**

b. Polar Protic Solvents and SN2

- A **polar protic solvent** contains either **–OH** or **-NH groups.**

Ex: CH_3OH (methanol), CH_3CH_2OH (EtOH = ethanol)

- Polar protic solvents are the **worst solvents in SN2. Why?**
- They are the **worst** solvents in SN2 because they **solvate** the incoming nucleophile thereby lowering its **ground state energy**. As a result, the energy of the transition state is **raised. Ea is larger** and the SN2 reaction is **slower**. The rate of the reaction is **decreased**.

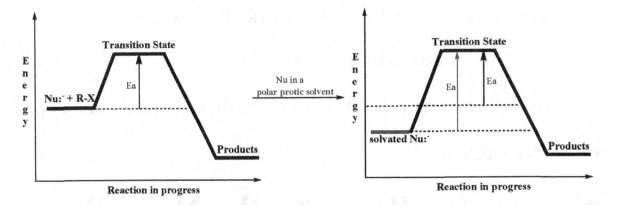

c. Polar Aprotic Solvents and SN2

- A **polar aprotic** solvent does not contain either **–OH** or **-NH** groups.
- Polar aprotic solvents are the **best solvents in SN2. Why?**
- They are the **best** solvents because they **do not solvate** the nucleophilic anion. Indeed, an **unsolvated** anion has a greater **nucleophilicity** since its ground state energy is **raised** in those solvents. **Ea is smaller.** As a result, the **energy of the transition** state is lower. The SN2 reaction is **faster.**

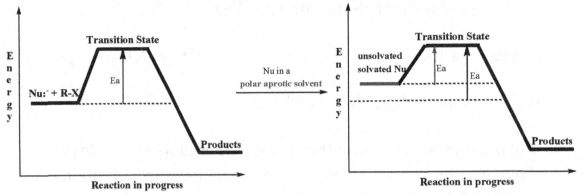

d. Some Polar Aprotic Solvents Used in SN2

- Acetone = CH_3COCH_3
- DMF = dimethylformamide

    $[(CH_3)_2NCHO]$
- Acetonitrile = CH_3CN or $CH_3C{\equiv}N$:

224

- Tetrahydrofuran =

- DMSO = dimethylsulfoxide

    $[(CH_3)_2SO]$

- HMPA = hexamethylphosphoramide

    $[((CH_3)_2N)_3PO]$

- The **increasing** order of solvent reactivity in SN2:

$CH_3OH < H_2O < DMF < CH_3CN < HMPA$

- See page _____. Do problems on page _____.

D. SN2: A SUMMARY

- Reaction is bimolecular = rate depends on concentrations of
- **both substrate and nucleophile.**
- Reaction proceeds in **one single step.**
- Reaction proceeds with **inversion of configuration** for chiral substrates (**Walden Inversion**).
- **No carbocation** intermediate.
- Best for **methyl and 1° alkyl** substrates.
- Never occurs with 3°, vinylic, and aryl substrates.
- **Negative nucleophiles** more reactive than neutral ones.
- Best leaving groups are the **conjugate bases** of strong acids.
- Best solvents are **polar aprotic** solvents.

E. THE SN1 REACTION

1. INTRODUCTION: EVIDENCE OF SN1 PATHWAY

Consider the following 2 reactions:

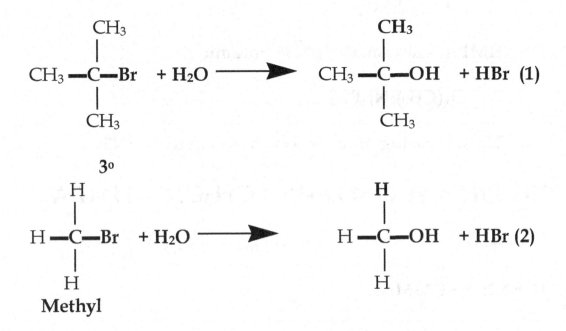

- **Observation: Reaction (1) is 1 million times faster than Reaction (2).**
- From SN2, we know that tertiary substrates are not good in SN2. Moreover, we know that neutral water is a poor nucleophile in SN2.

- **Conclusion: Reaction (1) cannot proceed via SN2. Therefore, there is an alternative substitution mechanism: SN1.**

S = Substitution

N = Nucleophilic

1 = unimolecular

2. THE KINETICS OF SN1

- The overall reaction is

$$Nu{:}^- \ + \ R{-}X \longrightarrow R{-}Nu + {:}X^-$$

$$\boxed{Rate = k[substrate]}$$

$$\boxed{Rate = k[R\text{-}X]}$$

- **Note: The reaction is first-order in [R-X] and is independent of the concentration of the nucleophile (the reaction is zero order in Nu). SN1 is said to be unimolecular.**

3. MECHANISM OF SN1

- An SN1 reaction proceeds in **2 steps**:
- **1st step: Slow formation of a carbocation intermediate**
- **2nd step: Fast attack** of the electrophilic carbocation by the nucleophile.

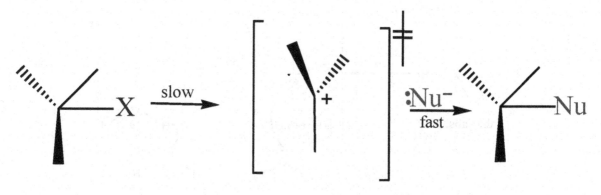

A planar carbocation intermediate

- **Note: The endergonic first step is the rate-determining step.**

$$R{-}X \xrightarrow{\text{slow}} [R^+] + :Nu^- \xrightarrow{\text{fast}} R{-}Nu + X^-$$

4. COMPARING SN1 AND SN2 ENERGY DIAGRAMS

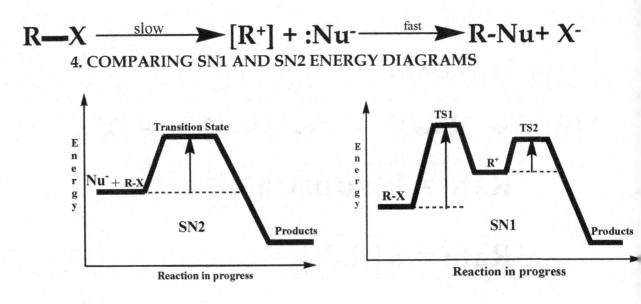

- See Fig. _____, page _____.

Ex: Show the mechanism of the reaction

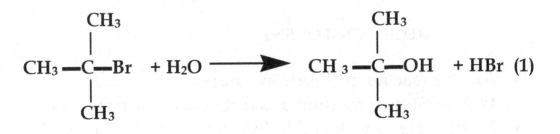

$$CH_3 - \underset{\underset{CH_3}{|}}{\overset{\overset{CH_3}{|}}{C}} - Br \ + H_2O \longrightarrow CH_3 - \underset{\underset{CH_3}{|}}{\overset{\overset{CH_3}{|}}{C}} - OH \ + HBr \ (1)$$

- Mechanism:

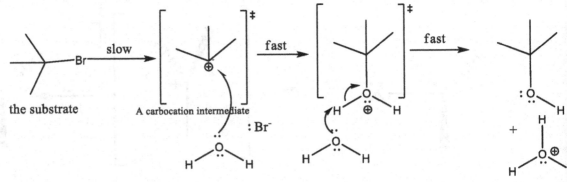

228

5. STEREOCHEMISTRY OF SN1 WHEN CHIRAL SUBSTRATES ARE USED

a. General Observation:

- **Recall: the carbocation intermediate is planar. Therefore, the C⁺ is sp² hybridized and has a vacant unhybridized p orbital.**
- **See Unit 8.**
- There are **2 ways** an **incoming nucleophile** can attack the carbocation intermediate
- **Back-side attack: results in an inversion of configuration** at the **stereogenic center.**
- **Front-side attack: results in retention of configuration.**

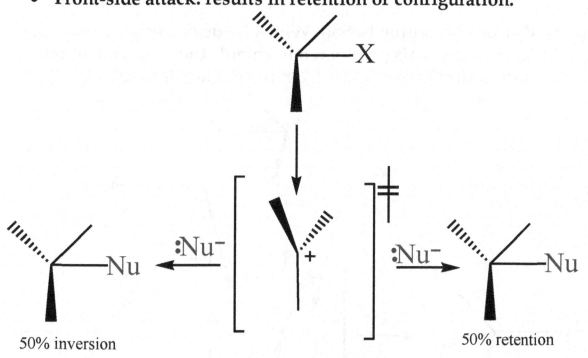

50% inversion   50% retention

A planar carbocation intermediate

- See page _____.

- **Conclusion: expect racemization: 50% R and 50% S.**

Ex:

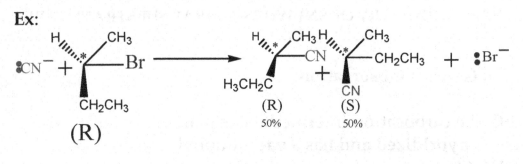

(R)

b. Deviation from Expectations

- In actuality, only few **chiral** substrates lead to complete racemization. In most cases, there is about 60% inversion and 40% retention of configuration. **Why?**

- **Reason: According to Saul Winstein, deviations are caused by ion-pairing. This phenomenon shields the front side of the carbocation intermediate from the nucleophilic attack.**

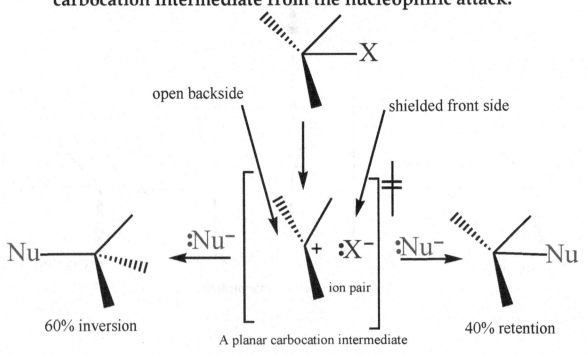

A planar carbocation intermediate

Ex:

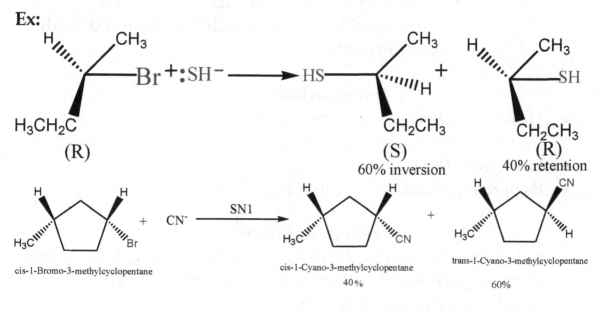

(R)   Br + :SH⁻ ⟶ HS— (S)   +   (R)

60% inversion   40% retention

cis-1-Bromo-3-methylcyclopentane

SN1

cis-1-Cyano-3-methylcyclopentane
40%

trans-1-Cyano-3-methylcyclopentane
60%

- Do problems on pages _____ – _____.

F. CHARACTERISTICS OF THE SN1 REACTION

1. INTRODUCTION:

- Like SN2, SN1 depends on the substrate, the leaving group, and the solvent.

- In general, factors that lower Ea (i.e., lowering of the energy of the transition state or raising the energy level of the ground state of the reactants) favor SN1 reactions.

2. EFFECT OF THE SUBSTRATE ON THE RATE OF SN1 REACTIONS
a. Introduction
- **Recall: SN1**

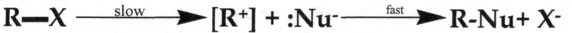

$$R\!-\!X \xrightarrow{\text{slow}} [R^+] + :Nu^- \xrightarrow{\text{fast}} R\text{-}Nu + X^-$$

- The more **stable** the carbocation intermediate, the **faster** the SN1 reaction.
- The increasing order of reactivity for common carbocations is

231

Methyl < 1º < allylic = benzylic < 2º < 3º.

b. Explanation of the order of stability observed with carbocations in SN1

i. Introduction
- There are **2 theories** on carbocation stability:

- **Inductive effects**
- **Hyperconjugation (See Unit 8).**

ii. Inductive Effect
- **Inductive effect** is the shifting of electrons in a bond due to the **electronegativities** of other atoms in the molecule. This is called **inductive stabilization.**
- Note: Metals and alkyl groups inductively donate electrons.
- **H does not donate electrons.**

- So the more alkyl groups present on the C^+, the more the electron density, the more stable the carbocation.

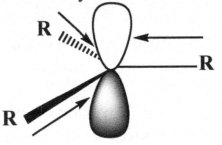

Ex:

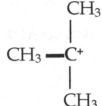

- 3 R groups ➔ inductive effect ➔ **more stable**

$$CH_3$$
$$|$$
$$H\!-\!C^+$$
$$|$$
$$H$$

232

- **No R groups → no inductive effect → less stable**
- **Note: allylic and benzylic carbocations are stabilized by resonance.**
- **Read pages** _____ - _____ .
- **Do problem on page** _____ .

iii. Hyperconjugation
- **Recall: A carbocation contains an sp², planar C⁺. See Unit 8.**
- **Hyperconjugation** is the overlap of the **vacant p** orbital on the carbocation with a properly oriented **σ (C-H)** bond on an adjacent C. The **more the alkyl substituents** on the C^+, the more the **possibilities** for hyperconjugation, the more stable the carbocation.

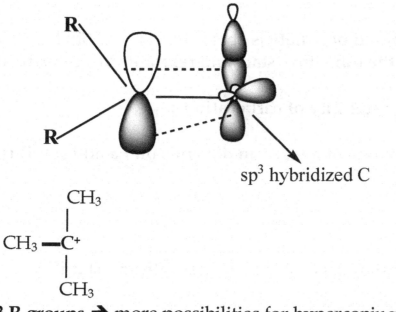

$$sp^3 \text{ hybridized C}$$

$$CH_3 - C^+ \begin{array}{c} CH_3 \\ | \\ \\ | \\ CH_3 \end{array}$$

- **3 R groups → more possibilities for hyperconjugation → more stable**

$$H - C^+ \begin{array}{c} H \\ | \\ \\ | \\ H \end{array}$$

- **No R groups** → no possibility for hyperconjugation → **less stable**
- See Fig. _____, page _____.

- See page _____.

c. The Hammond Postulate (or Hammond-Leffler postulate)

- Read pages _____ - _____.

- Read page _____ about SN1 applications to biological systems.

- The **Hammond postulate** is about the relationship between the **energy of the transition state and reactants or products in a reaction.**
- **Review the stability of carbocations:**

- **Recall: The rate of a reaction depends on Ea and stability depends on ΔG^o.**

i. Introduction

- **Note: Hammond's postulate relates rate and stability.**

- Hammond postulated that the transition state of a reaction resembles the structure of the substance to which it is **closer in energy**. There are **2 cases:**

- **Endergonic reactions**

- **Exergonic reactions**

ii. Endergonic Reactions ($\Delta G^o > 0$)

- The **transition state** resembles the **products** in structure since it is closer to products in energy.

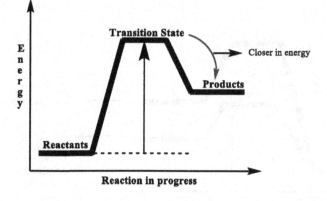

iii. Exergonic Reactions ($\Delta G^o < 0$)

- **The transition** state resembles the **reactants** in structure since it is closer to reactants in energy.

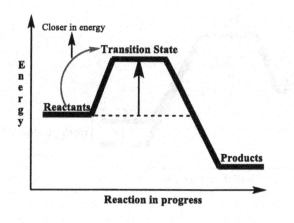

- See Fig. on page _____.

235

iv. Application of Hammond Postulate to Competitive or Parallel Reactions

- **Endergonic reactions: Different Eas (Ea1>Ea2)**

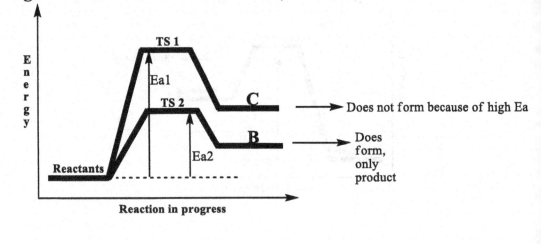

- See Fig. _____, page _____.

- Note: The more stable product forms faster in an endergonic reaction.

- **Exergonic reactions: same Eas; Both products observed**

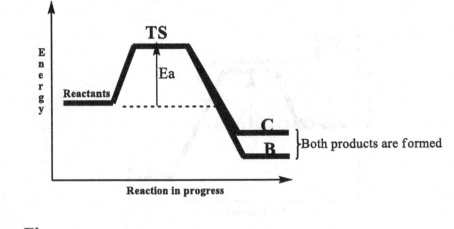

- See Fig. _____, page _____.

- Note: The more stable product may or may not form faster in an exergonic reaction.

d. Carbocation Formation and SN1

- Consider the 2 **endergonic** reactions in the formation of the 2 possible carbocations from propane:

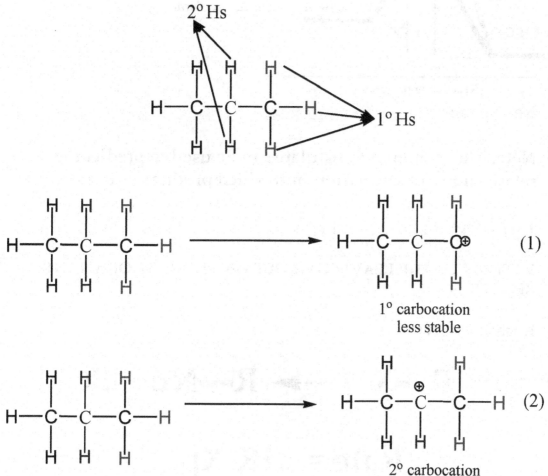

1° carbocation
less stable

(1)

2° carbocation
more stable

(2)

- Reaction 2 is faster than reaction 1 since a secondary carbocation is more stable and forms faster than a 1° carbocation. In actuality Reaction 2 is observed as **expected**.

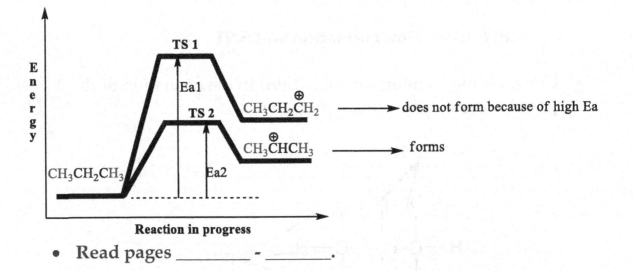

- Read pages _____ - _____.

- Note: The Hammond postulate can be used to predict the relative rates of 2 reactions and which product is formed.

- Do Problem _____, page _____.

3. EFFECT OF THE LEAVING GROUP ON THE RATE OF AN SN1 REACTION

- Recall: In SN1

$$Nu:^- \; + \; R{-}X \longrightarrow R{-}Nu + :X^-$$

$$\boxed{Rate = k[R\text{-}X]}$$

- Note: The rate of SN1 depends on the leaving group.

- Like in SN2, the best leaving groups in SN1 reactions are the **conjugate bases of strong acids**. The increasing order of reactivity is as follows:

$$H_2O < OH^- < Cl^- < Br^- < I^- = TosO^-$$

- Read page _____.

4. THE NUCLEOPHILE

- **Recall in SN1**

$$Nu{:}^- \ + \ R{-}X \longrightarrow R{-}Nu + {:}X^-$$

$$\boxed{Rate = k[R\text{-}X]}$$

- **Note:** The rate of SN1 <u>does not</u> depend on the nucleophile. In other words, the nature of the nucleophile is not that important. However, experiments have shown that SN1 reactions are favored with weakly, neutral nucleophiles.

5. THE SOLVENT

a. Dielectric Polarization (DP) and Solvent Polarity

- **DP** is the measure of the polarity of a solvent and its ability to solvate a carbocation.

Ex:

solvent	water	EtOH	Acetone	DMSO	hexane	formic acid
DP	80.4	24	21	49	1.9	59

- **Note:** Polar protic solvents have high DPs.
  polar aprotic solvents have low DPs.

b. Solvent Effect in SN1

- SN1 is favored in **protic solvents (high DP)** because these solvents **lower** the energy of the transition state by **solvating** the carbocation intermediate. In other words, they have a stabilizing effect. So they stabilize the carbocation intermediate.

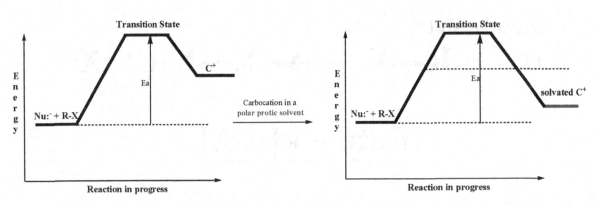

- **Note: SN1 reactions not favorable in polar aprotic solvents.**

G. SN1: A SUMMARY

- Reaction is **unimolecular** = rate depends **only** on the concentration of the substrate.
- Reaction proceeds in **two** steps; the **first step** is rate determining.
- A **carbocation intermediate is formed.**
- In general, reaction proceeds with **racemization** for chiral substrates
- Best for **3°, allylic, and benzylic** substrates.
- **Never occur with methyl and primary substrates due the unstable carbocation.**
- Rate **does not** depend on the concentration of the nucleophile.
- SN1 reactions favored with weakly, **neutral** nucleophiles
- Best leaving groups are the **conjugate bases** of strong acids
- Best solvents are **protic polar solvents**

H. DECISION-MAKING: SN1 OR SN2?

1. INTRODUCTION:
- In general, the mechanism of a nucleophilic substitution reaction depends on **4 factors**:
- **Substrate: 1°, 2°, 3°**
- **Nucleophile: strong or weak**
- **Leaving group: good or poor**
- **Solvent: protic or aprotic**

2. THE SUBSTRATE = MOST IMPORTANT FACTOR

a. 1° ➜ SN$_2$ only. No SN1.

Ex: CH_3Br, CH_3CH_2Cl

b. 3° ➜ SN1 only. No SN2

Ex: $(CH_3)_3CCl$

c. 2° ➜ both SN1 and SN$_2$ possible. Use other factors to choose.

Ex: $(CH_3)_2CHCl$

- See page _____.

- Read pages _____ - _____.

3. THE NUCLEOPHILE

a. Strong Nucleophiles (have negative charges)

- Strong nucleophiles **favor SN2.**

Ex: OH^-, CN^-, RS^-, SH^-, SeH^-, F^-, Cl^-, I^-, Br^-, RO^-

b. Weak Nucleophiles (neutral)

- Neutral, weak nucleophiles **favor SN1**.

Ex: H_2O, NH_3, ROH, RNH_2, H_2S

- See mechanism on page_____.

- Do problems on page _____.

Ex:

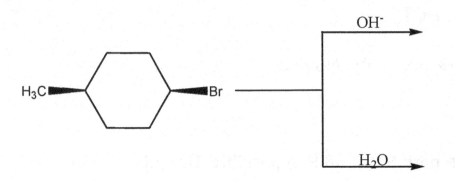

4. THE LEAVING GROUP

- **Good leaving groups enhance the rates of both SN1 and SN2.**

- For instance, for alkyl halides, **rate increases** as follows:

R-F < R-Cl < R-Br < R-I

Ex: CH_3Br, CH_3Cl

- SN2 ➔ faster for CH_3Br

5. THE SOLVENT

a. Polar protic solvents favor SN1 ➔ more stable carbocation by solvation

Ex: H_2O, ROH

b. Polar aprotic solvents favor SN2 ➔ free nucleophile; Nu not solvated

Ex: DMSO, acetone

- Do Problems on page _____.

- **Exercise**: Complete the following reactions and state the mechanism (SN1 or SN2) by which each one proceeds. Explain your choice

a.

b.

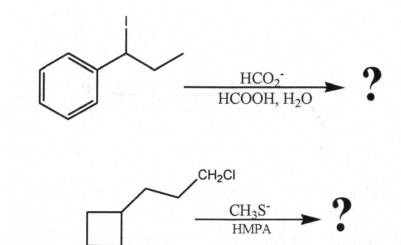

c.

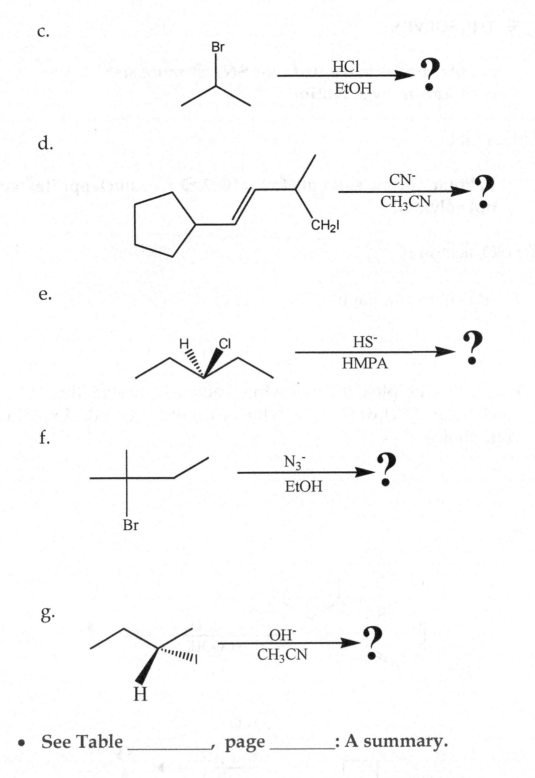

d.

e.

f.

g.

- **See Table _____, page _____: A summary.**

- **Read about non reactivity of vinyl and aryl halides, page ___.**

244

I. A WORD ABOUT ORGANIC SYNTHESIS

1. INTRODUCTION

$$? \longrightarrow \textbf{Products}$$

2. SN2 AND ORGANIC SYNTHESIS OF VARIOUS COMPOUNDS

- SN2 can be used to add a functional group to R. The general equation of the reaction is:

$$Nu{:}^- \ + \ R{-}X \longrightarrow R{-}Nu + :X^-$$

where $Nu{:}^-$ = OH^-, ^-OR, $RCOO^-$, CN^-, $HC{\equiv}C{:}^-$, N_3^-, NH_2^-, SH^-.

3. SUBSTITUTION PATTERNS OF SOME NEUTRAL (WEAK) NUCLEOPHILES (H-Nu)

Nucleophile (H-**Nu**)	Nucleophilic Part (**-Nu**)
H_2O	**-OH**
ROH	**-OR**
HCl	**-Cl**
NH_3	**-NH₂**
RSH	**-SR**
RCOOH	**-OOCR**
RNH_2	**-NHR**
HCN	**-CN**
H_2S	**-SH**

In general: SN2

$$H\text{-}Nu \ + \ R\text{-}X \longrightarrow R\text{-}Nu + HX$$

Ex:

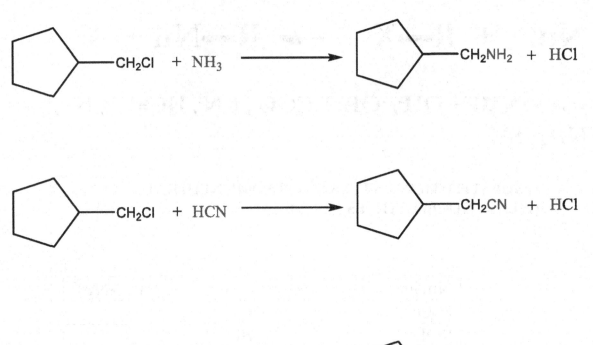

4. SN2 SYNTHESIS OF SOME FUNCTIONAL GROUPS

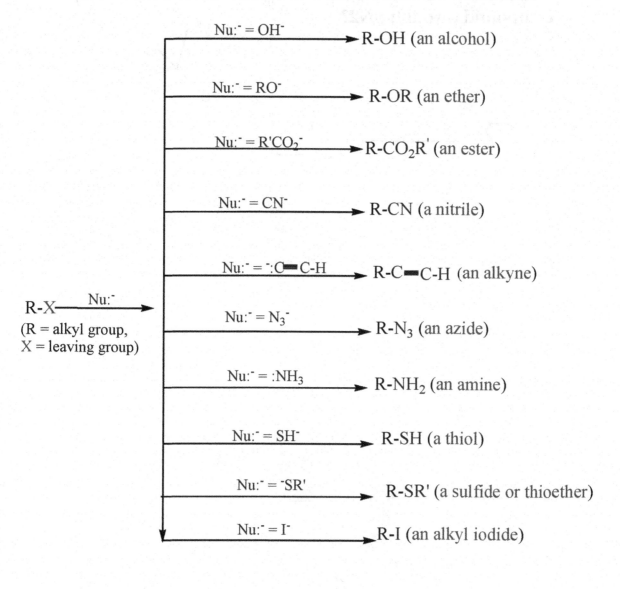

- See Table _____, page _____.

- **Read about the two-step synthesis of aspirin on page _____.**

- **Exercise: Please, find the appropriate substrate and nucleophile that can be used to synthesize the following compound through SN2?**

- See problems on page _____.

- See Key Concept, pages _____ - _____.

OCHEM UNIT 11: NUCLEOPHILIC ELIMINATION REACTIONS

A. INTRODUCTION: ELIMINATION vs. SUBSTITUTION: A REVIEW

- **Substitution Reactions: No Base Required. See Unit 10**

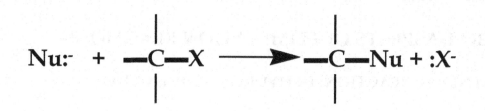

Ex: saturated product

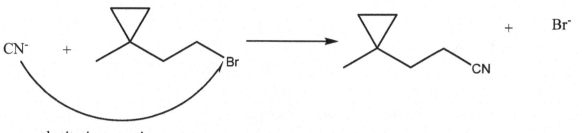

a substitution reaction

- **Elimination Reactions: A Base Is Required**

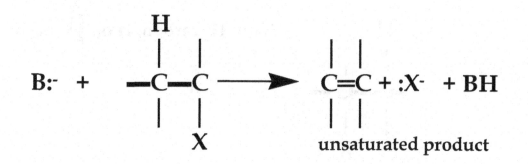

unsaturated product

249

Ex:

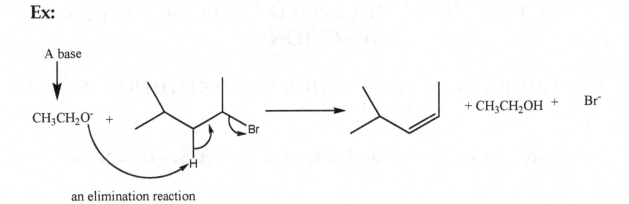

A base

$CH_3CH_2O^-$ +

→ + CH_3CH_2OH + Br^-

an elimination reaction

B. GENERAL ASPECTS OF ELIMINATION REACTIONS

1. GENERAL REACTION: DEHYDROHALOGENATION

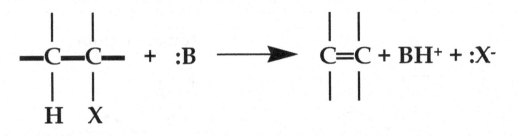

$$-\overset{|}{\underset{|}{C}}-\overset{|}{\underset{|}{C}}- \ + \ :B \longrightarrow \overset{|}{\underset{|}{C}}=\overset{|}{\underset{|}{C}} + BH^+ + :X^-$$

H   X

2. α AND β CARBONS

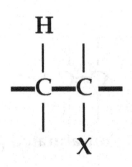

X on α carbon, H on β carbon

250

Ex:

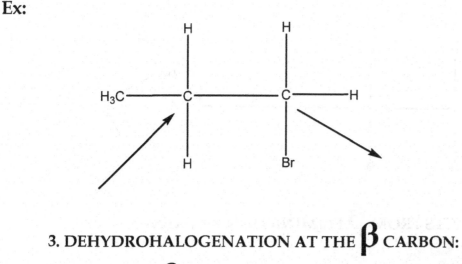

3. DEHYDROHALOGENATION AT THE β CARBON:

β ELIMINATION

The reaction is:

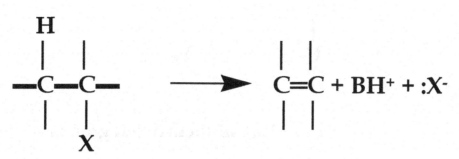

An alkene

where **:B = base = OH⁻, RO⁻, NH₂⁻, RS⁻, RC≡C:⁻**

- **Note:**

- **The α carbon loses the –X.**

- **The β carbon loses the –H.**

- **The double bond forms between the α and β carbons.**

- See Table _____, page _____ for common bases used.

- Do Problem _____, page _____.

Ex:

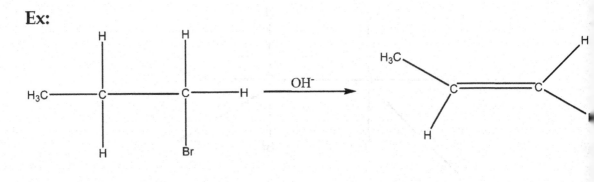

4. PRODUCTS FROM β ELIMINATION = ALKENES

a. Introduction

- There are **4 classes of alkenes: monosubstituted, disubstituted, trisubstituted, and tetrasubstituted.**
  - i. Monosubstituted: one R group

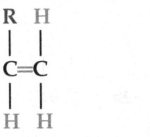

        Ex:

  - ii. disubstituted: two R groups

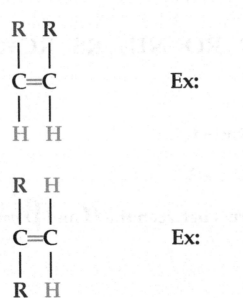

        Ex:

or

        Ex:

iii. trisubstituted: three R groups

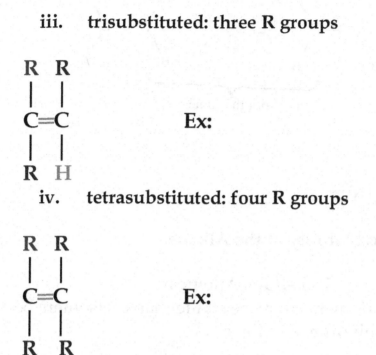

Ex:

iv. tetrasubstituted: four R groups

Ex:

- Do Problem _____, page _____.

b. The C=C Bond

- The C-C double bond is very rigid. Therefore, it **cannot be rotated**. This gives rise to two types of isomers when there are **two different groups on each carbon of the double bond: cis and trans. The** *cis* **and** *trans* **isomers are diastereomers.**
- **Note: If the same substituent is present on any of the 2 carbons of the double bond, there is no possibility of** *cis-trans* **isomerism.**

Ex:

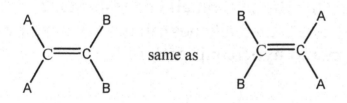

same as

253

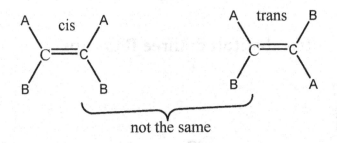

not the same

- Do problems on page _____ .

- Read pages _____ - _____ .

c. Stability of the Alkenes

i. cis-trans Alkenes

- **Trans** isomers are **more stable** than cis isomers because of **steric hindrance in the cis.**

Ex:

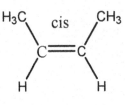

less stable

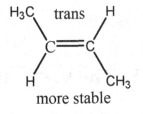

more stable

ii. Substituted Alkenes

- The stability of alkenes increases with **increasing substitution;** the more **R groups** present on the double bond carbons, the more **stable the alkene.**
- **Monosub.<disubst<trisubst.<tetrasubstituted.**

iii. Theories behind alkene stability: 2

- The stability of substituted alkenes can be explained by **inductive effect and hyperconjugation.** See OCHEM I UNITS 8 and 10.

- The R groups **stabilize** an alkene by **inductively** donating electrons to the carbon (s) of the double bond.

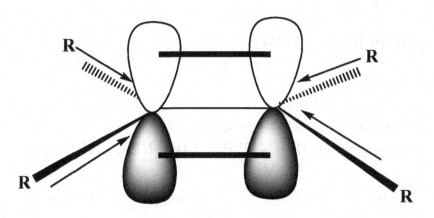

inductive effect

- Another reason for alkene stability is **hyperconjugation**, the **stabilizing** interaction between the empty antibonding π MO on the C=C bond and a filled σ MO on a C—H bond on an adjacent substituent. The more the substituents, the higher the possibilities for hyperconjugation, the more stable the alkene.

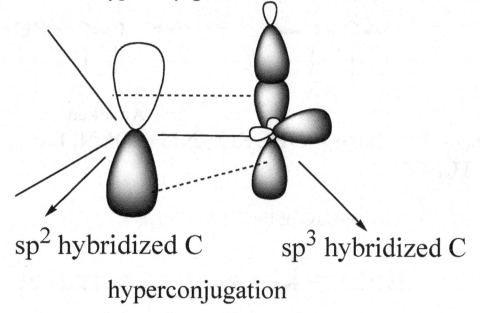

sp^2 hybridized C   sp^3 hybridized C

hyperconjugation

5. KINDS OF ELIMINATION REACTIONS

- There are 2 types of β elimination reactions: E1 and E2

C. THE E2 ELIMINATION REACTION

1. INTRODUCTION

- E2 is the most common elimination reaction.
- It occurs when **a strong base (DBN, DBU, OH⁻, RO⁻, NH₂⁻) is used.**
- See page _____ for IUPAC names of DBN and DBU.
- An **alkene** is produced.

2. THE KINETICS OF E2

- The E2 reaction proceeds as follows:

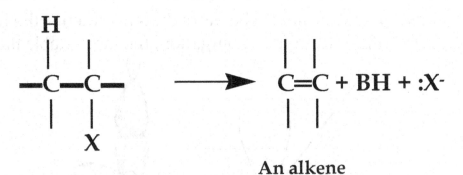

An alkene

where **:B⁻ = base = OH⁻, RO⁻, NH₂⁻, DBN, DBU, (CH₃)₃CO⁻**

- The rate law of the E2 reaction is:

$$\boxed{\text{Rate} = k[\text{base}][\text{substrate}]}$$

$$\boxed{\text{Rate} = k[\text{B}^-][\text{substrate}]}$$

- Note: The reaction is bimolecular. Therefore, the rate of the reaction depends on both base and substrate. The reaction is a second-order reaction.

- # E = Elimination
- # 2 = bimolecular

Ex:

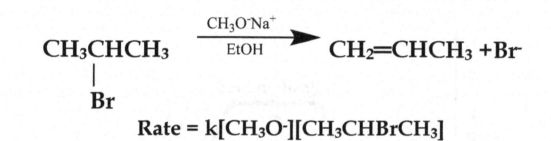

$$\text{CH}_3\text{CHCH}_3 \xrightarrow[\text{EtOH}]{\text{CH}_3\text{O}^-\text{Na}^+} \text{CH}_2{=}\text{CHCH}_3 + \text{Br}^-$$
$$|$$
$$\text{Br}$$

$$\text{Rate} = k[\text{CH}_3\text{O}^-][\text{CH}_3\text{CHBrCH}_3]$$

3. MECHANISM OF E2

- Similarly to SN2, the E2 reaction occurs in **one single step** with **concerted bond breaking and bond forming. No carbocation intermediate** is involved.

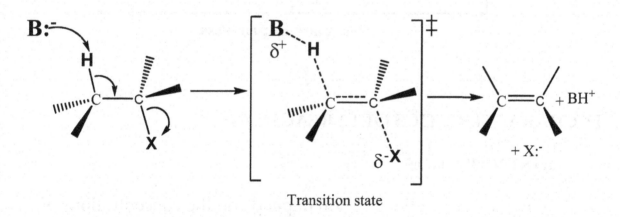

Transition state

257

Ex:

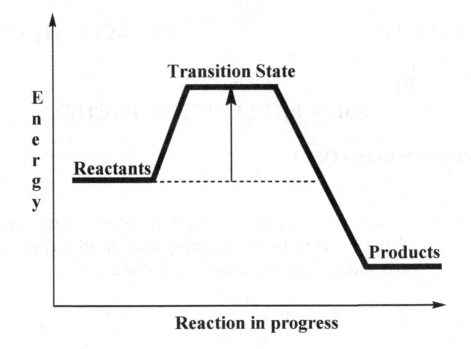

Transition state

4. THE ENERGY DIAGRAM OF THE E2 REACTION

- See Fig. _____, page _____.

D. CHARACTERISTICS OF E2 REACTIONS

1. INTRODUCTION

- The rate of the reaction depends on the concentrations of the substrate, the base, the leaving group, and the solvent.

- Read pages _____ – _____.

2. EFFECT OF THE SUBSTRATE ON E2

- The increasing order of reactivity is: $1^o < 2^o < 3^o$

- **The transition state is stabilized by the formation of a double bond and is also stabilized by inductive effect.**

- Read page _____. Do all related problems.

3. EFFECT OF THE BASE ON E2

- The base is **very important** in E2.
- **Recall:**

$$\boxed{\text{Rate} = k[B^-][\text{substrate}]}$$

- The rate of E2 increases with the strength of the base.

- **Common bases used: OH-, RO-, DBN, DBU, NH$_2$-, (CH$_3$)$_3$CO-,**
- See page _____.

4. EFFECT OF THE LEAVING GROUP ON E2

- As in SN2, the best leaving groups are the conjugate bases of strong acids. **The better the leaving group, the faster the reaction. The increasing order of reactivity is**

$$-F^- < Cl^- < Br^- < I^-$$

5. EFFECT OF THE SOLVENT ON E2

- Like in SN2, E2 is **faster in polar aprotic solvents (acetone, DMSO, etc.)**

- Do Problems _____, page _____ and _____, on page _____.

6. THE ZAITSEV RULE

- In **β** elimination reactions, the **more highly substituted alkene product is the major product.**

Ex:

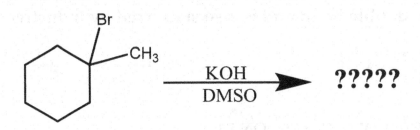

7. STEREOCHEMISTRY OF E2

a. Some Definitions

- **A reaction is regioselective when only one possible constitutional isomer is formed. The most stable isomer is formed.**

- **E2 is regioselective.**

Ex:

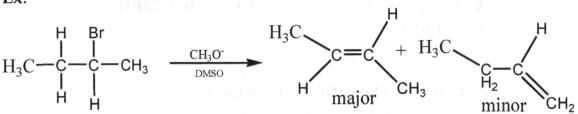

- **A reaction is stereoselective when one stereoisomer is mostly formed out of 2 or more possibilities.**

- **E2 is stereoselective.**

Ex:

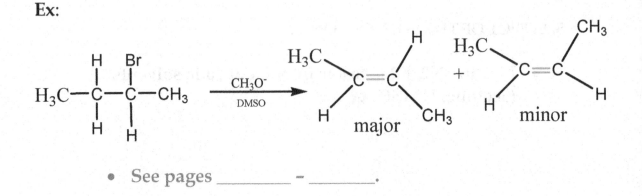

- See pages _____ - _____.

b. Stereochemistry of E2

i. Syn Periplanar (Eclipsed) – Antiperiplanar (Staggered)

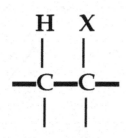

- **H and X on same side = syn periplanar (eclipsed)**

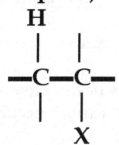

- **H and X on opposite sides = anti periplanar (staggered)**

- Note: E2 occurs with antiperiplanar geometry.

Ex:

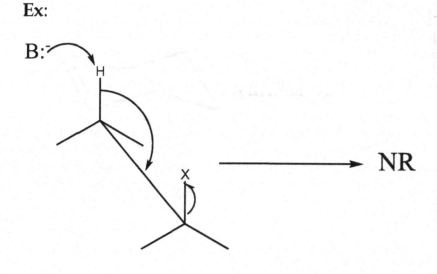

NR

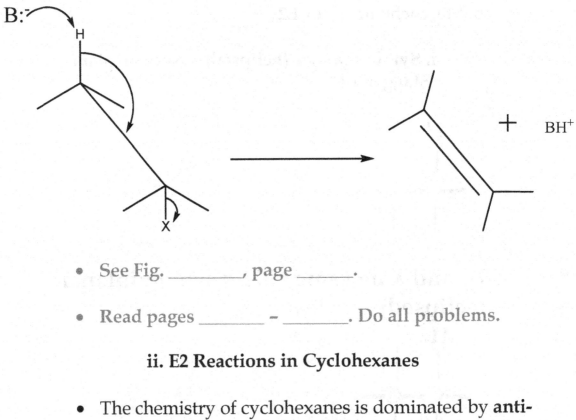

- See Fig. _____, page _____.

- Read pages _____ – _____. Do all problems.

ii. E2 Reactions in Cyclohexanes

- The chemistry of cyclohexanes is dominated by **anti-periplanar reactions**. Indeed, **E2 cannot occur** if the leaving group or H is **equatorial**.

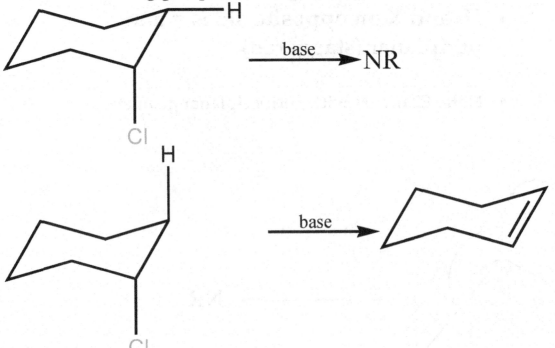

- Read pages _____ - _____. Do Problems on page _____.

8. THE KINETIC DEUTERIUM ISOTOPE EFFECT

a. Introduction
- **The Bond C-D is stronger than the C-H bond. So C-H is much easier to break than C-D.**

b. The Reactions

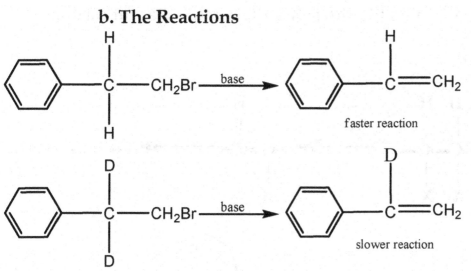

- **Conclusion: The C-H bond is broken in the rate-determining step.**

9. SUMMARY ON E2 REACTIONS
- E2 is a **second-order** reaction; its rate depends on both the concentrations of the base and the substrate; the reaction is **bimolecular**.
- E2 is a **one-step reaction** (no carbocation).
- E2 proceeds with **antiperiplanar geometry**.
- E2 occurs when strong bases such as OH^-, RO^-, NH_2^-, DBU, DBN are used in **polar aprotic solvents**.
- E2 reaction faster for 3° substrates.
- E2 faster with better leaving groups.
- E2 shows a kinetic deuterium isotope effect.

- See Table _____, page _____. Do all problems.

10. APPLICATION OF E2 TO ALKYNE SYNTHESIS USING VICINAL AND GEMINAL DIHALIDES

a. Introduction
- The dehydrohalogenation requires **strong bases** such as NH_2^- ($NaNH_2$) or $KOC(CH_3)_3$ in DMSO.

b. Vicinal Dihalides: X on separate adjacent carbons

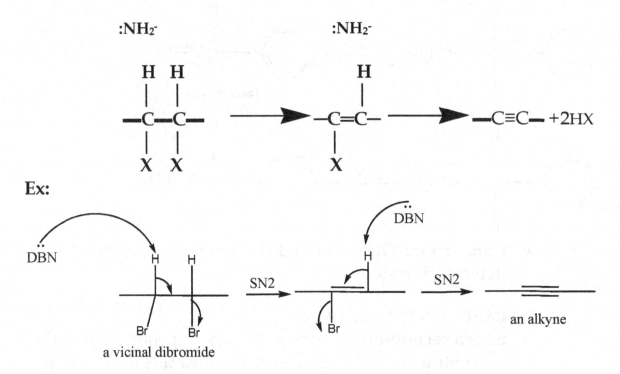

Ex:

a vicinal dibromide

an alkyne

c. Geminal Dihalides: X on the same carbon

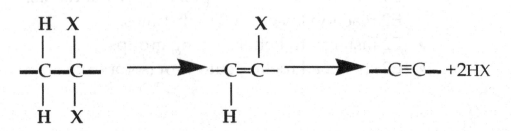

Ex:

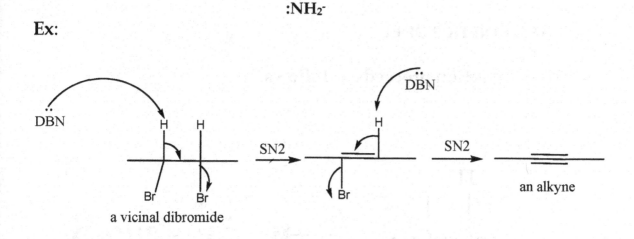

a vicinal dibromide   SN2   SN2   an alkyne

E. THE E1 REACTION

1. INTRODUCTION

- E1 is the second type of elimination reactions.
- It usually occurs along SN1 when a reaction is carried out in a **polar protic solvent** (H_2O, ROH) with a **nonbasic (neutral) nucleophile.**

Ex:

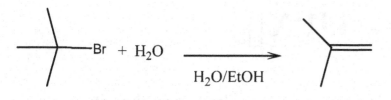

H_2O = non basic nucleophile

H_2O/EtOH = polar protic solvent

2. THE KINETICS OF E1

- The **E1 reaction** proceeds as follows:

:B

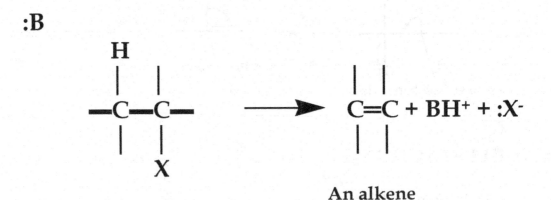

An alkene

where **:B = base = H_2O, ROH**

- The rate law of the E1 reaction is:

$$\boxed{\textbf{Rate = k[substrate]}}$$

$$\boxed{\textbf{Rate = k[RX]}}$$

- Note: The reaction is unimolecular. Therefore, the rate of the reaction depends only on the substrate. The reaction is a first-order reaction.

- **E = Elimination**
- **1 = unimolecular**

Ex:

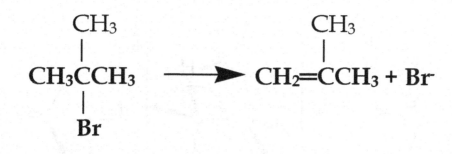

$$\text{Rate} = k[(CH_3)_3CBr]$$

3. MECHANISM OF E1

- E1 is a **two-step mechanism reaction (like SN1)**

  a. Formation of a carbocation (rate-determining step).

  b. A fast abstraction of a β H by a base.

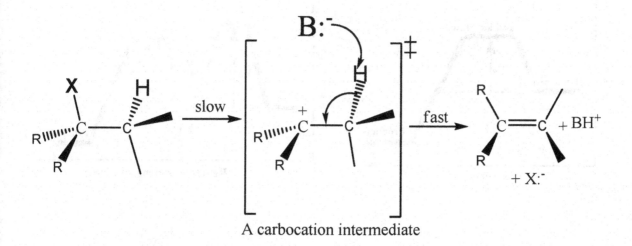

A carbocation intermediate

Ex:

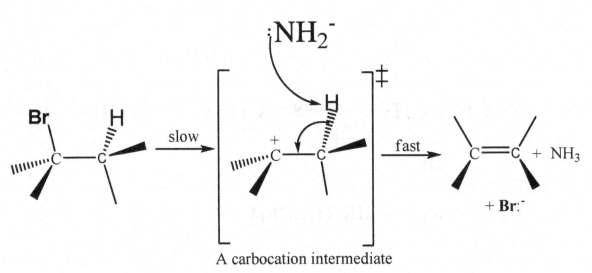

A carbocation intermediate

- **Note: SN1 and E1 are in competition when a nonbasic nucleophile is used.**

- Do all problems.

F. COMPARATIVE ENERGY DIAGRAMS OF E2 AND E1

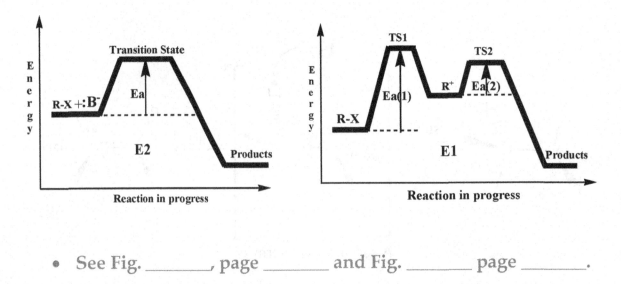

- See Fig. _____, page _____ and Fig. _____ page _____.

G. CHARACTERISTICS OF E1

1. THE SUBSTRATE

- Like in E2, the increasing order of the E1 reaction is: **1º < 2º < 3º**.
- This is due to the order of stability of the **carbocation** intermediate.
- The reaction is **regioselective** since the **most substituted alkene** product is formed.

Ex:

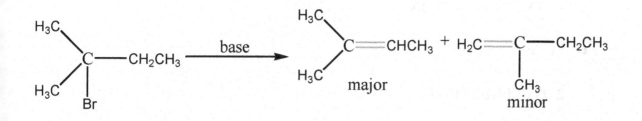

2. THE BASE

- The rate of E1 is **independent** of the concentration of the base. However, **weaker bases such** H_2O and ROH favor E1.

3. THE SOLVENT

- As in SN1, E1 is favored in **polar protic solvents**.

4. THE LEAVING GROUP

- The E1 reaction is faster when better leaving groups are involved.

5. SUMMARY OF E1

- E1 is **unimolecular: 1st-order reaction**
- E1 proceeds in **2 steps**
- Formation of a **carbocation**
- E1 occurs (with SN1) when a **nonbasic nucleophile** is used in a **polar protic solvent.**
- E1 reaction is faster with 3° substrate.
- **E1 does not require antiperiplanar geometry**
- **E1 does not show a deuterium isotope effect**

- See Table _____, page _____.

H. E1 vs. E2

1. INTRODUCTION

- Is the reaction E1 or E2?

2. THE SUBSTRATE

- The substrate has the same effect on both E1 and E2. Therefore, the substrate **cannot** be used to differentiate between E1 and E2.

3. THE BASE

- Strong bases (OH⁻, RO⁻, NH₂⁻) favor E2.
- Weak neutral bases (ROH, H₂O) favor E1.

- **Note: E1 does not occur with 1° alkyl halides since the primary carbocation is not stable. Do examples.**

I. DECISION MAKING: WHICH IS IT?

1. INTRODUCTION

Given:

$$Nu:^- + R{-}X \longrightarrow \text{????}$$

Will RX undergo substitution or elimination? **It's difficult to say!!!!!!!!!!!!!!!** There is no easy answer because the reaction can give a mixture of products.

2. GENERAL OBSERVATIONS

- **Good nucleophiles that are weak bases** favor substitution over elimination. Some weak bases that are good nucleophiles are: **I^-, HS^-, CN^-, Br^-, and CH_3COO^-.**

Ex:

$$CH_3CH_2CH_2Br + HS^- \longrightarrow CH_3CH_2CH_2SH + Br^-$$

- **Bulky and strong bases ($(CH_3)_3CO^-$, OH^-, NH_2^-, RO^-, RS^-)** favor elimination over substitution.

Ex:

$$CH_3CH_2CH_2Br + OH^- \longrightarrow CH_3CH{=}CH_2 + Br^-$$

- **Higher temperatures favor elimination over substitution because more molecules can overcome high elimination activation energies . Furthermore, at higher temperatures, there is an increase in entropy (Recall: $\Delta G^o = \Delta H^o - T\Delta S^o$).**

271

Ex:

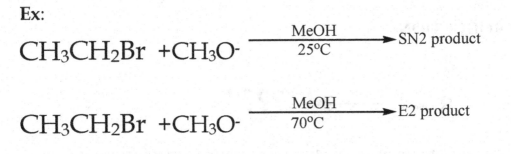

$$CH_3CH_2Br + CH_3O^- \xrightarrow[25°C]{MeOH} \text{SN2 product}$$

$$CH_3CH_2Br + CH_3O^- \xrightarrow[70°C]{MeOH} \text{E2 product}$$

3. OTHER CRITERIA TO CONSIDER IN PRODUCT PREDICTION

- First, classify the substrate as a **1° , 2° , or a 3°.** Next, identify the base or nucleophile as **nucleophile only, base or bulky base only, strong base and nucleophile, or neutral nucleophile and base.**

4. REACTIONS WITH TERTIARY SUBSTRATES

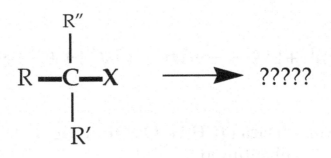

- **SN2 never occurs → steric hindrance**

- **With strong or bulky bases only ((CH₃)₃CO⁻ K⁺, NH₂⁻, DBU, DBN, LDA, (CH₃)₃CLi), E2 only occurs.**

Ex:

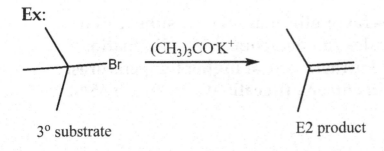

3° substrate   E2 product

- **With neutral, weak nucleophiles or bases (RSH, H₂O, NH₃, ROH, RNH₂, H₂S, RCOOH), both SN1 and E1 occur. A mixture of products is obtained.**

Ex:

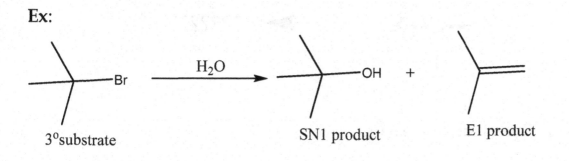

3°substrate   SN1 product   E1 product

- **With nucleophiles only (Cl⁻, HS⁻, I⁻, Br⁻, RS⁻, RCOO⁻), only SN1 occurs.**

Ex:

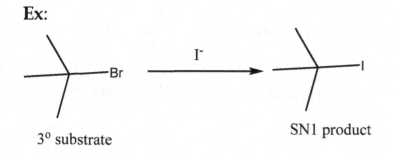

3° substrate   SN1 product

- **With strong bases and nucleophiles (OH⁻, RO⁻), E2 only occurs.**

Ex:

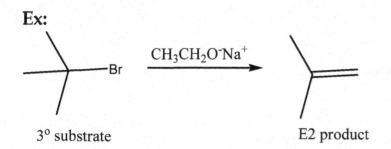

3° substrate   E2 product

5. REACTIONS WITH SECONDARY SUBSTRATES

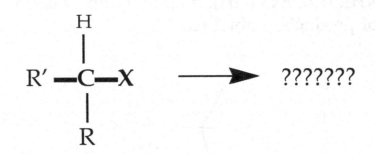

- All mechanisms occur.

- With strong bases and nucleophiles (OH⁻, RO⁻), both SN2 and E2 occur → get a mixture of products

Ex:

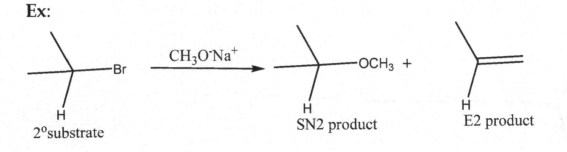

- With bases only or bulky bases (($CH_3)_3CO^-K^+$, DBU, DBN, NH_2^-), only E2 occurs.

Ex:

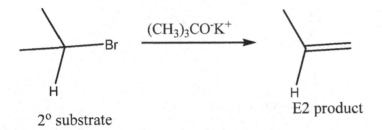

- **With nucleophiles only (CH$_3$COO⁻ , I⁻, CN⁻), SN2 and SN1 occur.**

Ex:

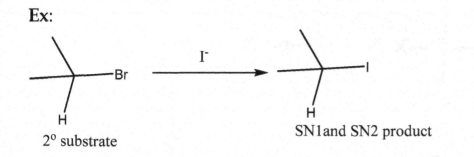

2° substrate

SN1 and SN2 product

- **With neutral nucleophiles or weak bases (H$_2$O, ROH), both SN1 and E1 occur➜ get a mixture of products.**

Ex:

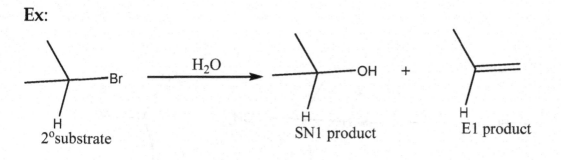

2° substrate

SN1 product

E1 product

- **With 2° benzylic and allylic substrates, SN1 and E1 can also occur.**

Ex:

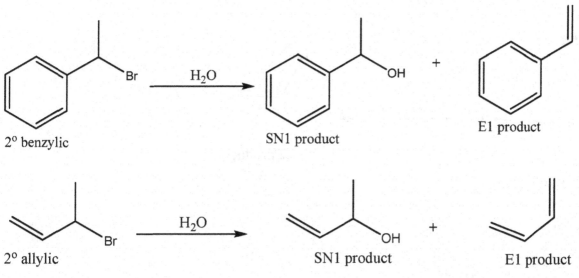

2° benzylic

SN1 product

E1 product

2° allylic

SN1 product

E1 product

6. REACTIONS WITH PRIMARY SUBSTRATES

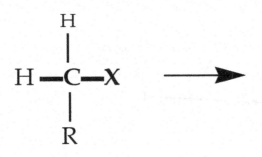

- **Never SN1 or E1 → unstable carbocation.**

- **With strong bases and nucleophiles only (OH⁻, RO⁻), both SN2 and E2 occur.**

Ex:

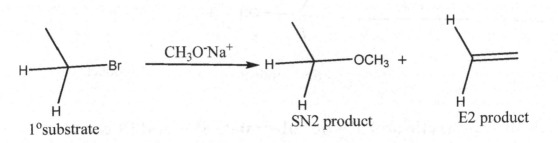

- **With bases and bulky bases only ((CH₃)₃CO⁻K⁺, NH₂⁻, DBU, DBN...), only E2 occurs.**

Ex:

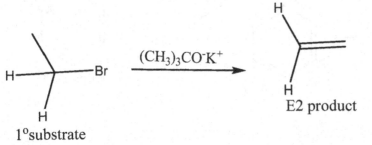

- With nucleophiles only (CH_3COO^- , I^-, CN^-), only SN2 occurs.

Ex:

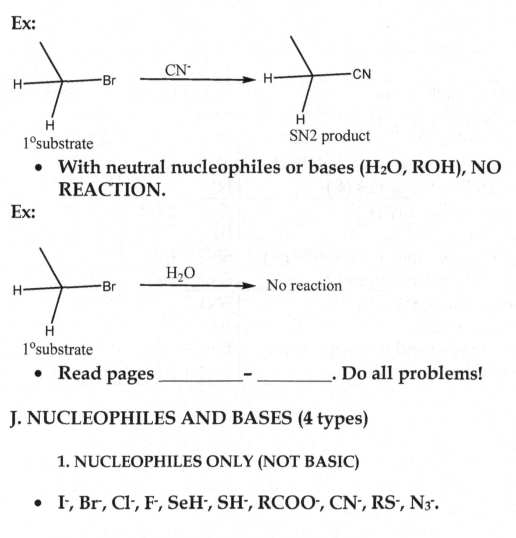

1°substrate

SN2 product

- With neutral nucleophiles or bases (H_2O, ROH), NO REACTION.

Ex:

1°substrate

- Read pages _____ – _____. Do all problems!

J. NUCLEOPHILES AND BASES (4 types)

1. NUCLEOPHILES ONLY (NOT BASIC)

- I^-, Br^-, Cl^-, F^-, SeH^-, SH^-, $RCOO^-$, CN^-, RS^-, N_3^-.

2. BASES ONLY (NOT NUCLEOPHILES)

- NH_2^-, DBU, DBN, $(CH_3)_3CO^-$ M^+ (M=Na, K, Li), $(CH_3)_3Li$, LDA (Lithium DiisopropylAmide)

3. STRONG BASES AND NUCLEOPHILES (BOTH)

- OH^-, RO^- (EtO^-, MeO^- ...), $RC\equiv C:^-$, Na^+, RLi.

4. NEUTRAL WEAK NUCLEOPHILES AND WEAK BASES

- H_2O, NH_3, ROH, RNH_2, H_2S, RSH, RCOOH.

K. SN1, SN2, E1, E2: A SUMMARY

- See pages _____ - _____. See Table _____, page _____
  for summary.

Substrate	Base or nucleophile	Mechanism	Never occurs
1°	-nucleophile only (1.) -base only (2.) -Strong base and nucleophile (3.) -neutral nucleophiles (4.)	SN2 E2 SN2 + E2 NR	SN1 E1
2°	-nucleophile only (1.) -base only (2.) -strong base and nucleophile (3.) -neutral nucleophiles (4.)	SN2 + SN1 E2 SN2 + E2 SN1 + E1	0
3°	-nucleophile only (1.) -base only (2.) -strong base and nucleophile (3.) -neutral nucleophiles (4.)	SN1 E2 E2 SN1 + E1	SN2

or

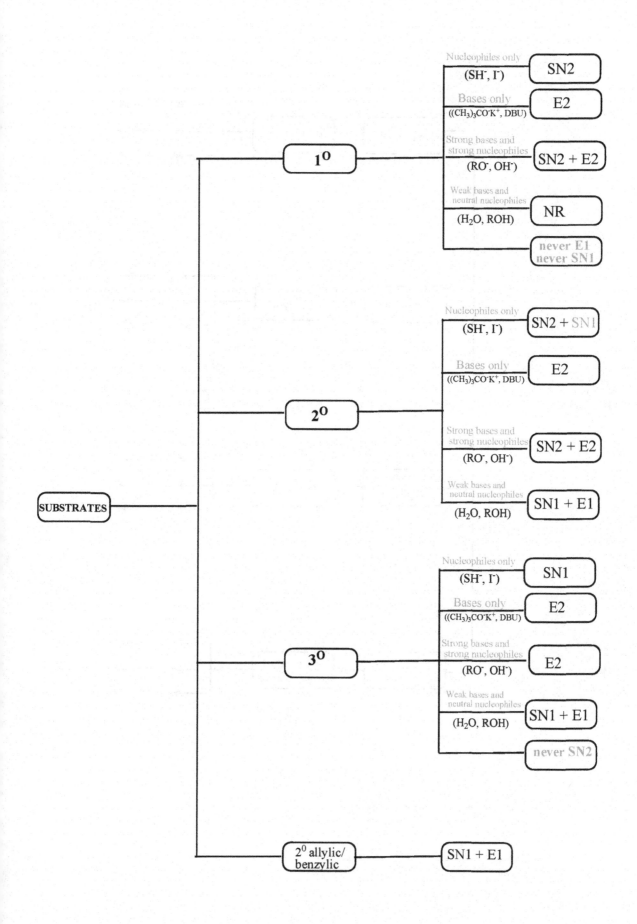

or

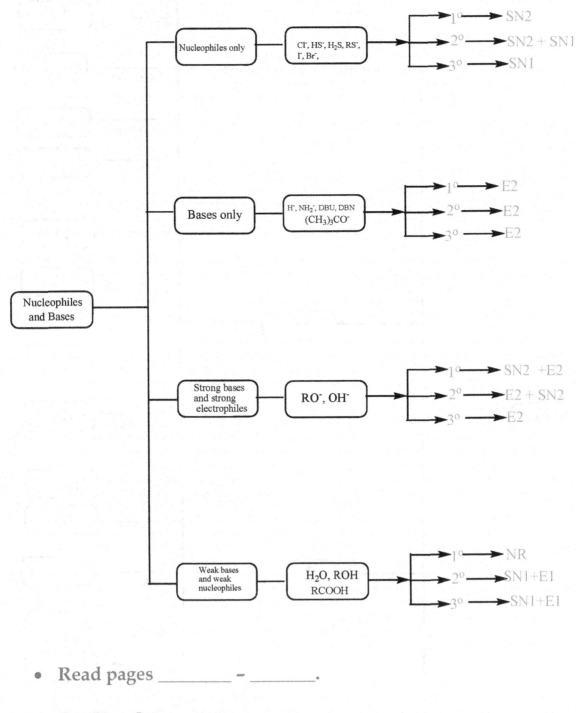

- Read pages _____ – _____.

- See Key Concepts on pages _____ – _____.

280

- **Exercise**: Complete the following reactions and state the mechanism (s) (E1, E2, SN1, SN2) that each is undergoing.

a.

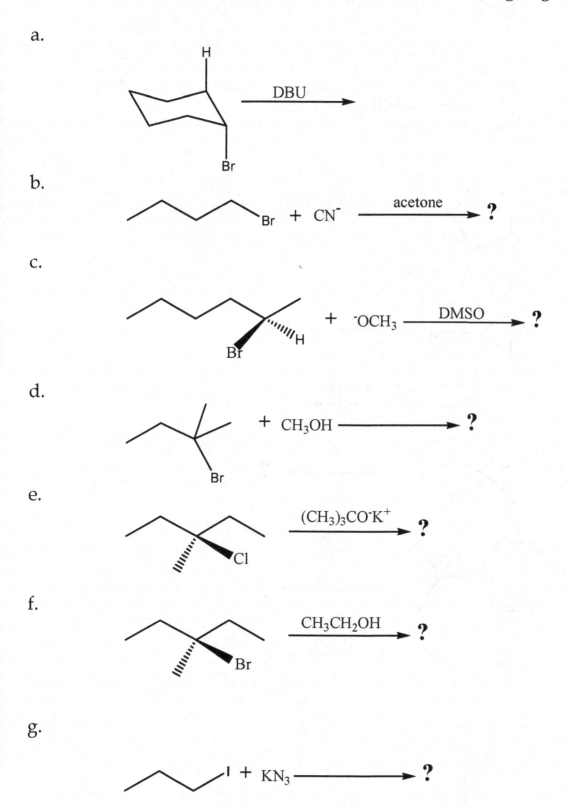

b.

c.

d.

e.

f.

g.

h.

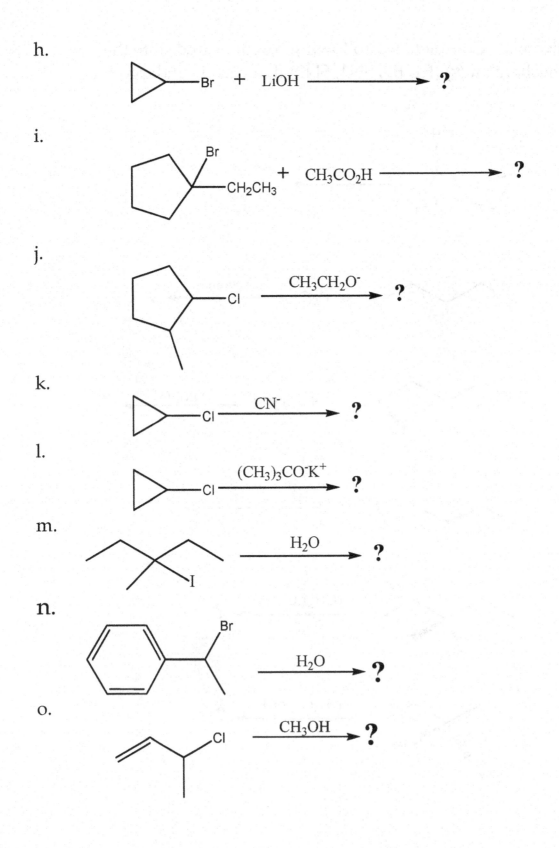

i.

j.

k.

l.

m.

n.

o.

OCHEM I UNIT 12: ALCOHOLS

A. GENERAL STRUCTURE OF ALCOHOLS

1. INTRODUCTION

- An alcohol is a compound that contains an **–OH** group. The general structure of an alcohol is

$$\mathbf{R\!-\!OH}$$

where R is an alkyl group.

Ex:

- **Phenols** have the general structure:

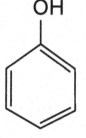

- **Enols** have the general structure:

$$\mathbf{R\!-\!\overset{\displaystyle OH}{\underset{\displaystyle |}{C}}\!=\!CH_2}$$

Ex:

- **Note: Alcohols can be thought as derivatives of water:**
H–OH.

2. CLASSIFICATION OF ALCOHOLS

i. Introduction

- There are **3 classes** of alcohols: **1°, 2°, 3°**.

ii. Primary Alcohols have 2 H Atoms on C Bearing OH

1° alcohol

Ex:

iii. Secondary Alcohols Have 1 H Atom on C Bearing OH

2° alcohol

Ex:

iv. Tertiary Alcohols Have No H Atoms on C Bearing OH

3° alcohol

Ex:

- **Read pages _____ – _____ ; Do problem on page _____.**

B. NOMENCLATURE OF ALCOHOLS

1. IUPAC NAMES

- Names of alcohols end in —ol. They are named as alkan —ol.

Ex:

$$CH_3CH_2OH \qquad \text{ethanol or ethan-1-ol}$$

- **Note: In naming alcohols, the parent chain is the longest carbon chain that contains the –OH group. Carbons in the parent chain are numbered from the end nearest the C bearing the OH group.**

Ex:

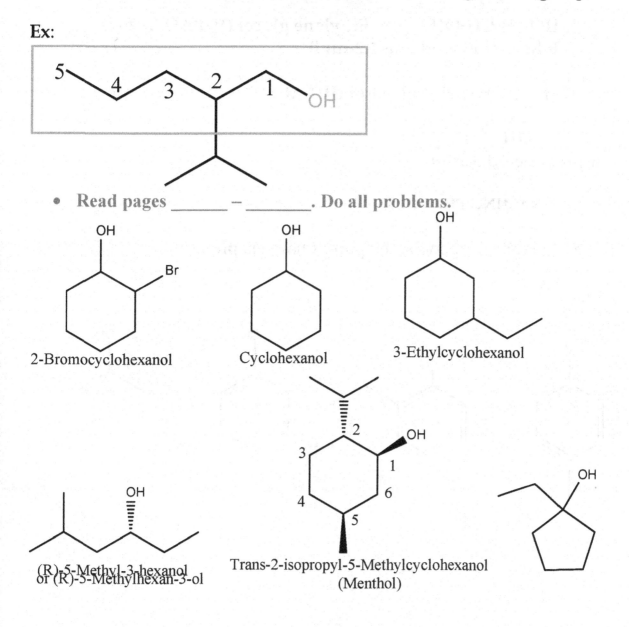

- **Read pages _____ – _____. Do all problems.**

2-Bromocyclohexanol

Cyclohexanol

3-Ethylcyclohexanol

(R)-5-Methyl-3-hexanol
or (R)-5-Methylhexan-3-ol

Trans-2-isopropyl-5-Methylcyclohexanol
(Menthol)

285

2. COMMON NAMES

- Alcohols can be also named as **alkyl alcohols**.

Ex:

CH_3CH_2OH Ethyl alcohol

CH_3OH Methyl alcohol

$H_2C=CHCH_2OH$ Allyl alcohol (IUPAC: 2-propen-1-ol)

$HOCH_2CH_2OH$ Ethylene glycol (IUPAC: 1, 2 – Ethanediol or ethane-1,2-diol)

$HOCH_2\,CHCH_2OH$ glycerol (IUPAC: 1,2,3-propanetriol)
|
OH
or propane-1,2,3-triol

3. NAMING PHENOLS

- In naming phenols, the parent name is **phenol.**

Ex:

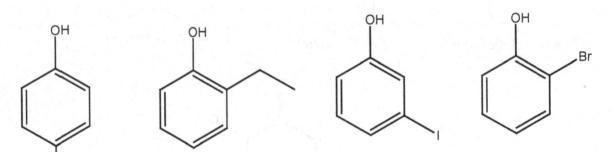

4. NAMING POLYFUNCTIONAL ALCOHOLS

a. Group Priorities

- When two or more different functional groups are present in a compound, functional group priorities are used. Group priorities are assigned based on the following Table (decreasing order of priority).

Priority order	Group	Ending of name as a priority	Name as a **non** **Priority group**
Carboxylic acid	**RCOOH**	*-oic acid*	*-carboxy*
Ester	**RCOOR'**	*-oate*	*-alkoxycarbonyl*
Amide	**RCONH₂**	*-amide*	*-amido*
Nitrile	**RCN**	*-nitrile*	*-cyano*
Aldehyde	**RCHO**	*-al*	*-oxo(=O) or formyl(-CHO)*
Ketone	**RCOR'**	*-one*	*-oxo*
Alcohol	**ROH**	*-ol*	*-hydroxy*
Amine	**RNH₂**	*-amine*	*-amino*
Alkene	**-C=C-**	*-ene*	*-alkenyl*
Alkyne	**-C≡C-**	*-yne*	*-alkynyl*
Alkane	**-C-C-**	*-ane*	*-alkyl*
Ether	**ROR'**	*-none*	*-alkoxy*
Halide	**R-X**	*-none*	*-halo*

b. Some Examples

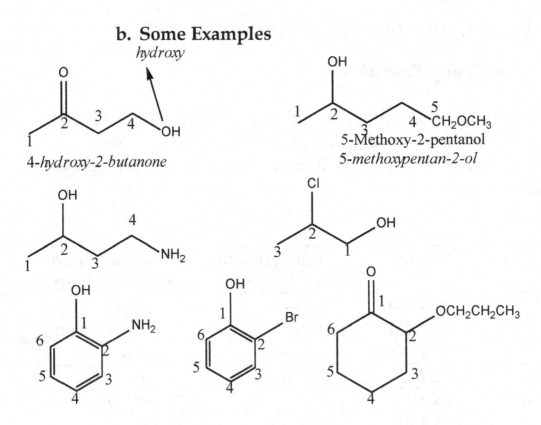

hydroxy

4-*hydroxy-2-butanone*

5-Methoxy-2-pentanol
5-*methoxypentan-2-ol*

5. NAMING ALCOHOLS AND OTHER COMPOUNDCONTAINING DOUBLE BONDS AND OTHER FUNCTIONAL GROUPS

- **The numbering follows the priority table. However, the parent chain is always named as an alkene. Some Examples**

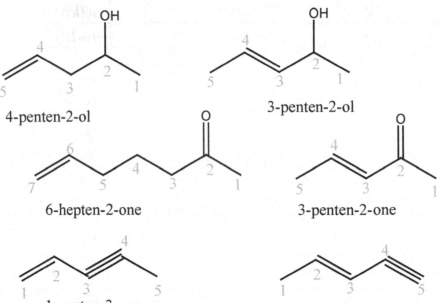

4-penten-2-ol

3-penten-2-ol

6-hepten-2-one

3-penten-2-one

1-penten-3-yne

2-penten-4-yne

288

C. PHYSICAL PROPERTIES OF ALCOHOLS

1. BONDING IN ALCOHOLS

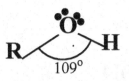

- Alcohols are OH-containing polar substances. Therefore, they have strong intermolecular bonds (VWF, DD, H bonds).

2. BP AND MP

- Alcohols have elevated BP and MP due the formation of H bonds between their molecules.

Ex:

$$CH_3CH_2CH_2OH \quad BP = 97\ ^oC$$

$$CH_3CH_2CH_2CH_3 \quad BP = -0.50\ ^oC$$

3. SOLUBILITY

- If the number of carbons in alcohol ≤ 5, then the alcohol is soluble in water. Otherwise, it is insoluble.
- All alcohols are soluble in organic solvents.

- See Table _____, page _____.

- Do problem on page _____.

D. IMPORTANT ALCOHOLS

- Read pages _____.

- most important alcohols are

Methanol: CH_3OH

2-propanol: $CH_3\,CHCH_3$

$|$

OH

Ethanol: CH_3CH_2OH

Ethylene glycol: $HOCH_2CH_2OH$

Estradiol and cholesterol

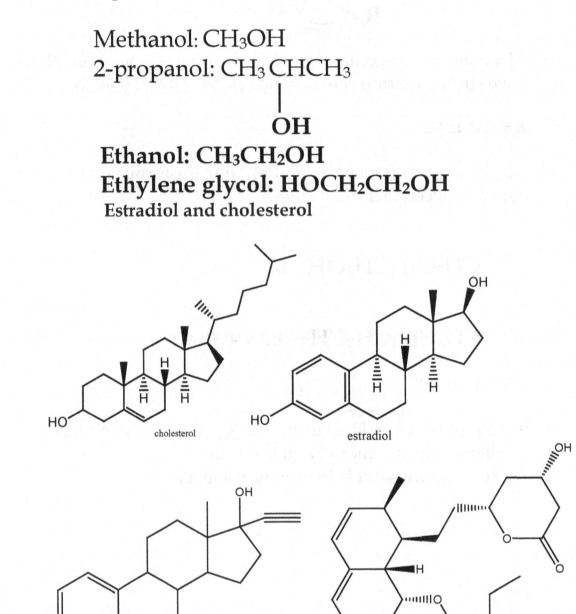

cholesterol

estradiol

mestranol

zocor

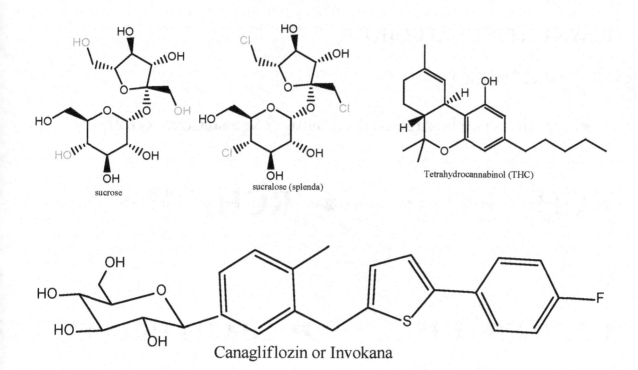

sucrose

sucralose (splenda)

Tetrahydrocannabinol (THC)

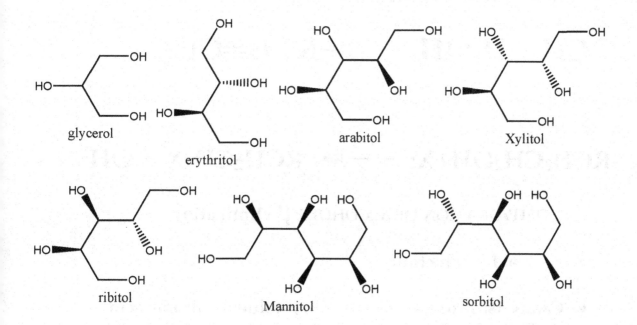

Canagliflozin or Invokana

- **Sugar alcohols = polyols or polyalcohols used as sweeteners**

glycerol

erythritol

arabitol

Xylitol

ribitol

Mannitol

sorbitol

E. SYNTHESIS OF ALCOHOLS

1. INTRODUCTION

- Alcohols can be prepared by using **SN$_2$ reactions** as follows:

$$RCH_2X + OH:^- \longrightarrow RCH_2OH + :X^-$$

Ex:

$$CH_3CH_2CH_2Br + OH:^- \longrightarrow CH_3CH_2CH_2OH + :Br^-$$

F. REACTIONS OF ALCOHOLS

1. GENERAL REACTIONS

- Alcohols can undergo both **β elimination and nucleophilic substitution** reactions.

$$RCH_2CH_2OH \longrightarrow RCH=CH_2$$

or

$$RCH_2CH_2OH + X:^- \longrightarrow RCH_2CH_2X + OH^-$$

2. DEHYDRATION OF ALCOHOLS: β elimination

a. Introduction

- Catalysts: strong acids : H_2SO_4, p-toluenesulfonic acid (TsOH), 85% H_3PO_4.

- **General Reaction:**

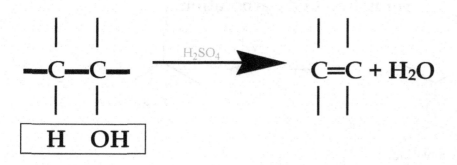

$$-\overset{|}{\underset{|}{C}}-\overset{|}{\underset{|}{C}}- \quad \xrightarrow{H_2SO_4} \quad \overset{|}{C}=\overset{|}{C} + H_2O$$

Ex:

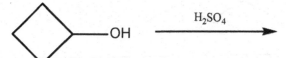

- Note: The increasing order of reactivity: $1^o < 2^o < 3^o$.
- 2° and 3° alcohols proceed via E1.
- 1° alcohols react thru **E2.**
- A double bond is formed between the α and β carbons.
- The reaction follows Zaitsev's rule.

- Do Problems on page _____ and page _____.

b. Mechanism I: 2° and 3° Alcohols: E1

i.Step 1: Formation of the Substrate

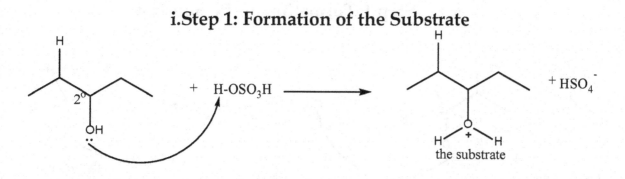

293

ii. Step 2: Loss of H₂O, a Good Leaving Group: Formation of a Carbocation

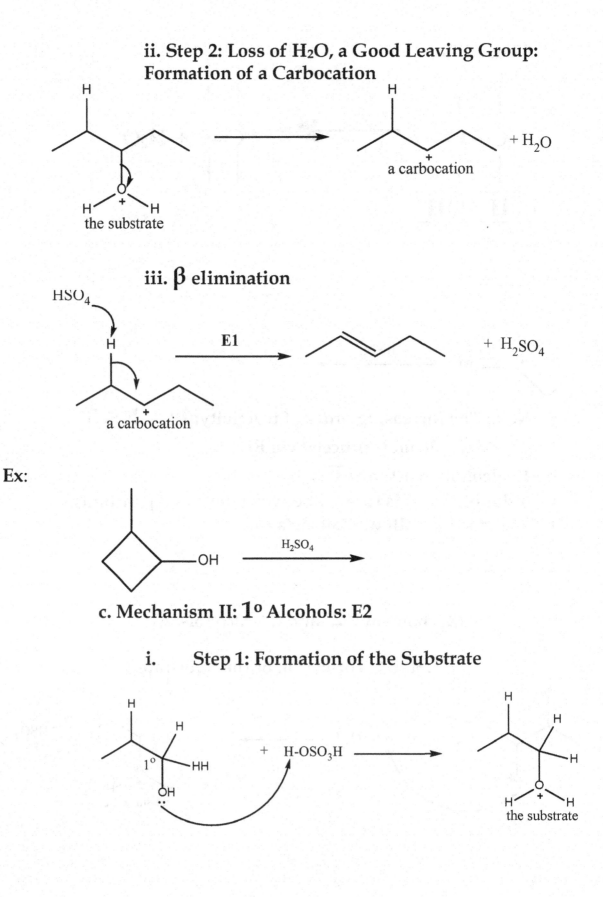

iii. β elimination

Ex:

c. Mechanism II: 1° Alcohols: E2

i.   Step 1: Formation of the Substrate

ii. Step 2: β Elimination and Loss of H₂O, a Good Leaving Group

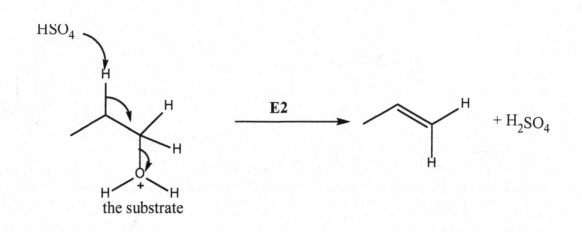

Ex:

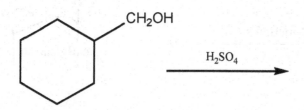

2. CARBOCATION REARRANGEMENTS IN THE DEHYDRATION OF ALCOHOLS: β Elimination

a. 1,2-Methyl Shift

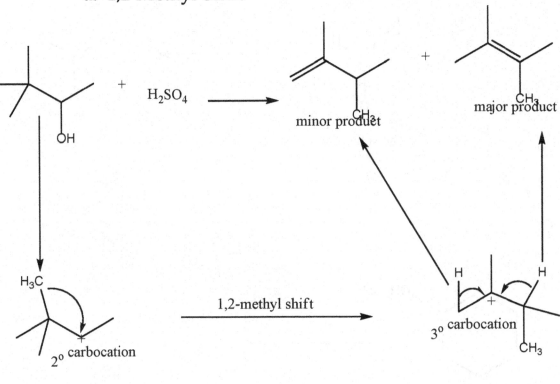

- See mechanism on page _____.

b. 1,2-Hydride Shift

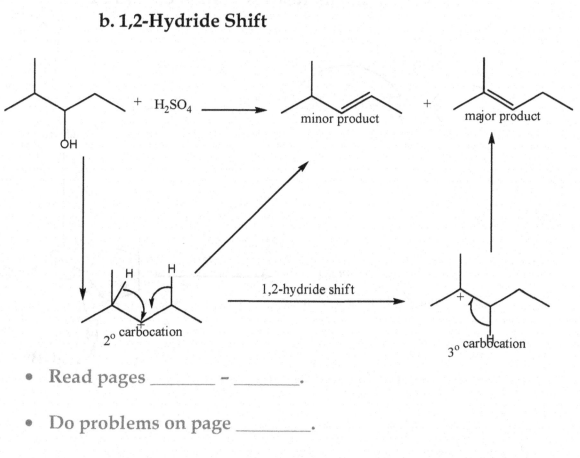

minor product   +   major product

1,2-hydride shift

$2°$ carbocation $3°$ carbocation

- Read pages _____ – _____.

- Do problems on page _____.

4. DEHYDRATION OF ALCOHOLS USING POCl₃ IN PYRIDINE

a. Reaction with POCl₃ = phosphorus oxychloride

- **General Reaction: Use DBU or pyridine. A Zaitzev product is obtained.**

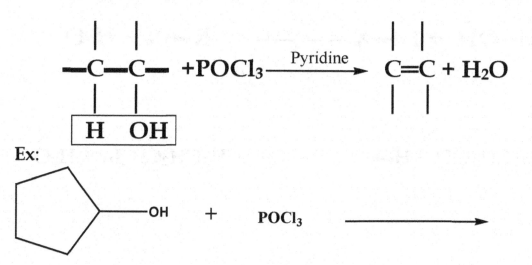

$$-\overset{|}{\underset{|}{C}}-\overset{|}{\underset{|}{C}}- + POCl_3 \xrightarrow{\text{Pyridine}} \overset{|}{\underset{|}{C}}=\overset{|}{\underset{|}{C}} + H_2O$$

$\boxed{H \quad OH}$

Ex:

OH   +   POCl₃ ⟶

b. Mechanism: Reaction Proceeds thru E2

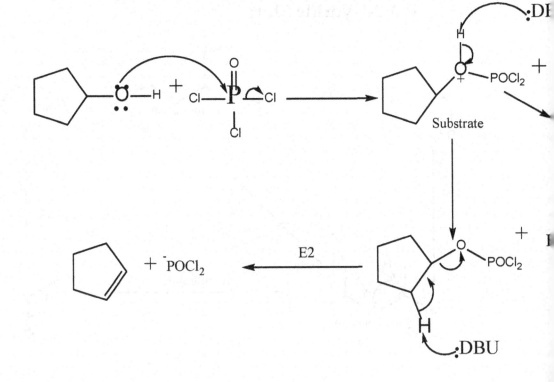

- See page _____.

- Read pages _____ – _____.

5. SUBSTITUTION REACTIONS OF ALCOHOLS WITH HX (X = Cl, Br, I): SYNTHESIS OF ALKYL HALIDES

a. General Reaction

$$R\text{—OH} + H\text{—X} \longrightarrow R\text{—X} + H_2O$$

Ex:

$$CH_3CH_2CH_2OH + H\text{—Br} \longrightarrow CH_3CH_2CH_2Br + H_2O$$

b. Mechanism

i. Introduction

- Primary alcohols react thru an **SN2** mechanism.
- Secondary and tertiary alcohols proceed thru **SN1**.

ii. 1° alcohols: SN2

- See page _____.

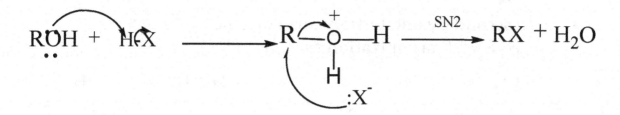

iii. 2° and 3° alcohols:SN1

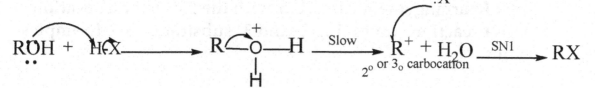

- **Note: The decreasing order of reactivity for substrates is :**
  $3^o > 2^o > 1^o$.
- **Increasing order of reactivity for HX: HCl<HBr<HI.**

- **Note: For HCl, a ZnCl$_2$ catalyst is required.**

- Read pages _____ – _____.

Ex:

$$CH_3CH_2OH + H\!-\!Cl \xrightarrow{\text{ZnCl}_2} CH_3CH_2Cl + H_2O$$

c. Stereochemistry of the Product for Chiral Alcohols

- With **primary alcohols,** the **SN2** reaction proceeds with **inversion of configuration as expected.**

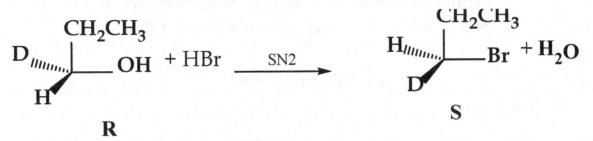

- With **secondary and tertiary alcohols,** the **SN1** reaction proceeds with **racemization as expected.**

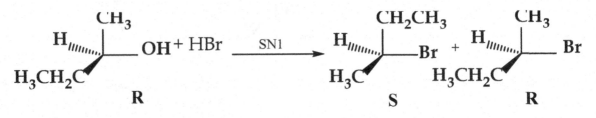

- **Note: Rearrangement can occur with the formed carbocation in SN1 reactions involving 2° and 3° substrates. See Examples below.**

- **Can have a Methyl shift:**

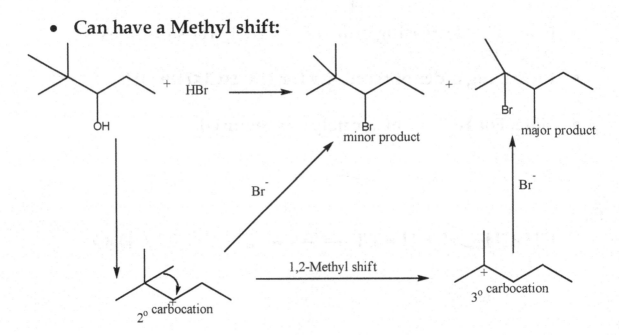

300

- **Can have a hydride shift:**

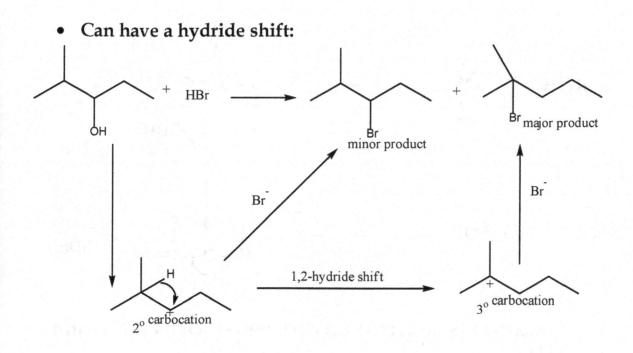

6. REACTIONS OF ALCOHOLS WITH THIONYL CHLORIDE IN PYRIDINE: $SOCl_2$

a. General Reaction: SN2

- The reaction occurs with **primary and secondary alcohols and methanol (CH_3OH).**

$$ROH + SOCl_2 \xrightarrow{\text{Pyridine}} RCl + SO_2 + HCl$$

Ex:

$$CH_3CH_2OH + SOCl_2 \xrightarrow{\text{Pyridine}} CH_3CH_2Cl + SO_2 + HCl$$

b. See reaction mechanism, page _____.

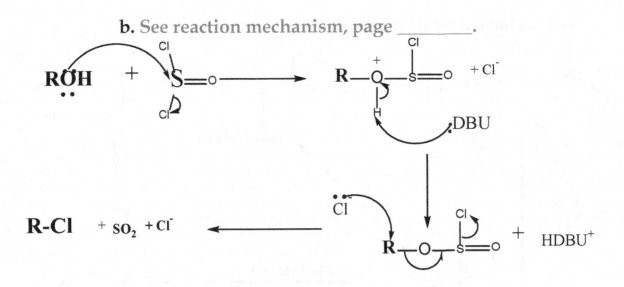

7. REACTIONS OF ALCOHOLS WITH PHOSPHORUS TRIBROMIDE : PBr_3

a. General Reaction: SN2

- The reaction occurs with **primary and secondary alcohols and methanol (CH_3OH) (the best alcohol in this reaction).**

$$ROH + PBr_3 \longrightarrow RBr + HOPBr_2$$

Ex:

$$CH_3CH_2OH + PBr_3 \longrightarrow CH_3CH_2Br + HOPBr_2$$

b. See reaction mechanism = SN2: page _____.

- See Table _____, page _____.
- Read pages _____ - _____.
- See Summary, page _____.
- Do problems on page _____.

$$ROH \,(1° \text{ and } 2°) + SOCl_2 \xrightarrow{\text{Pyridine}} RCl + SO_2 + HCl$$
$$ROH \,(1° \text{ and } 2°) + PBr_3 \longrightarrow RBr + HOPBr_2$$
$$ROH \,(1° \text{ and } 2°) + PCl_3 \longrightarrow RCl + HOPCl_2$$

8. FORMATION OF TOSYLATES FROM ALCOHOLS: TsO⁻

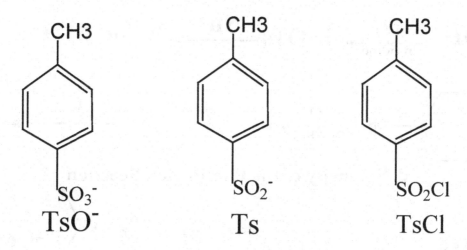

$$TsO^- \qquad\qquad Ts \qquad\qquad TsCl$$

- The tosylate ion (a very good leaving group)

a. Formation of the Tosylates from Alcohols in Pyridine

$$RCH_2OH + TsCl \xrightarrow{\text{Pyridine}} RCH_2OTs$$

Ex:

$$CH_3CH_2OH + TsCl \xrightarrow{\text{Pyridine}} CH_3CH_2OTs$$

b. General Reaction with Tosylates with Nucleophiles

$$RCH_2OTs + :Nu^- \longrightarrow RCH_2Nu + OTs^-$$

Ex:

$$CH_3CH_2OTs + CN:^- \longrightarrow CH_3CH_2CN + OTs^-$$

- Note: The reaction is SN2 if a strong nucleophile (RO⁻, CN⁻) is used.

- However, the reaction is E2 if a bulky or strong base ((CH₃)₃CO⁻ K⁺) is used.

c. Summary of the Substitution Reaction

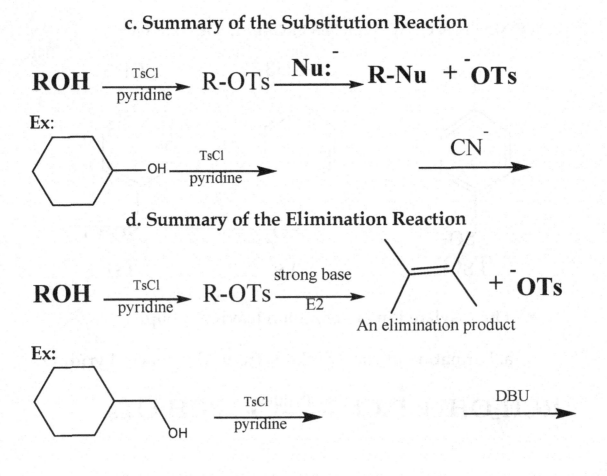

ROH $\xrightarrow[\text{pyridine}]{\text{TsCl}}$ R-OTs $\xrightarrow{\text{Nu:}^-}$ R-Nu + $^-$OTs

Ex:

d. Summary of the Elimination Reaction

ROH $\xrightarrow[\text{pyridine}]{\text{TsCl}}$ R-OTs $\xrightarrow[\text{E2}]{\text{strong base}}$ + $^-$OTs

An elimination product

Ex:

9. THE PINACOL REARRANGEMENT REACTION

a. Introduction

- This reaction converts vicinal diols (or glycols) to aldehydes and ketones using sulfuric acid as a catalyst.

304

b. The General Reaction For Symmetrical Diols

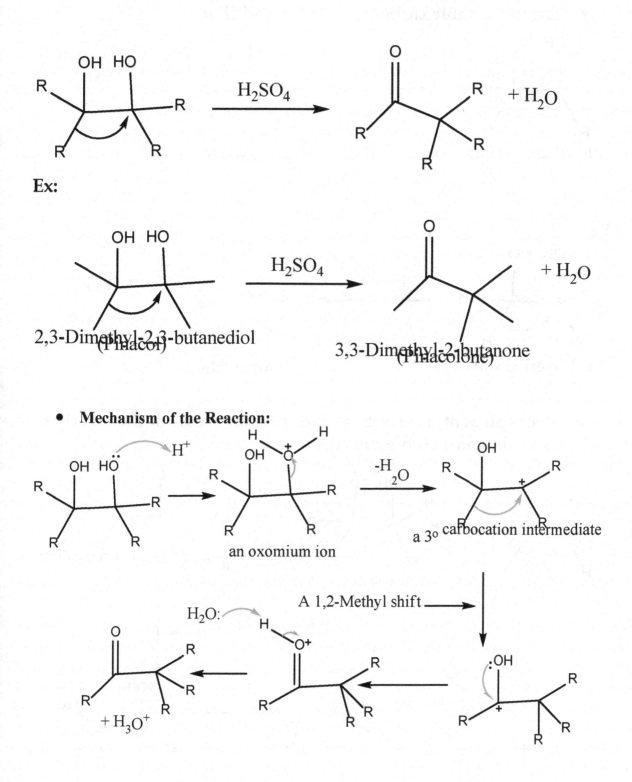

Ex:

2,3-Dimethyl-2,3-butanediol
(Pinacol)

3,3-Dimethyl-2-butanone
(Pinacolone)

- **Mechanism of the Reaction:**

an oxomium ion

a 3° carbocation intermediate

A 1,2-Methyl shift

+ H_3O^+

c. The General Reaction For Unsymmetrical Diols

- **The most stable carbocation is formed first.**

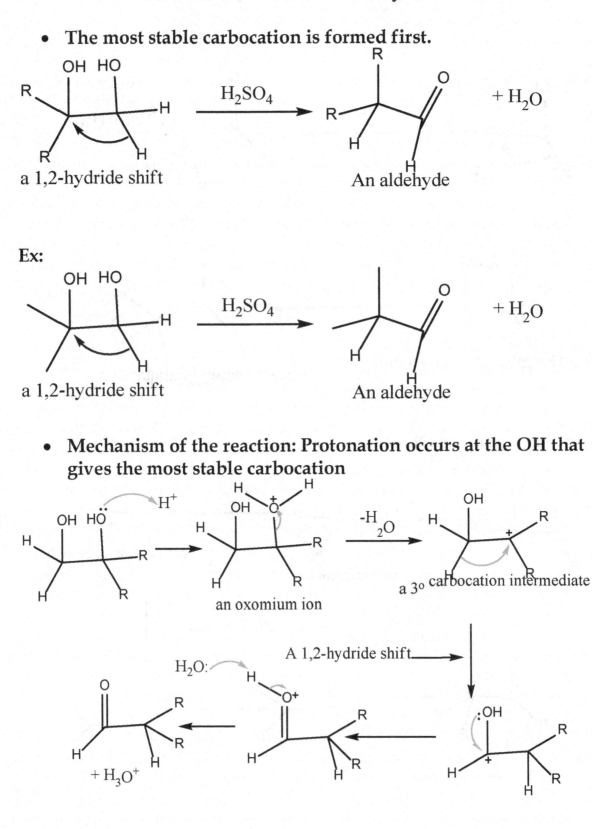

a 1,2-hydride shift

An aldehyde

$+ H_2O$

Ex:

a 1,2-hydride shift

An aldehyde

$+ H_2O$

- **Mechanism of the reaction: Protonation occurs at the OH that gives the most stable carbocation**

an oxomium ion

a 3° carbocation intermediate

$-H_2O$

A 1,2-hydride shift

$H_2O:$

$+ H_3O^+$

10. REACTION OF ALCOHOLS WITH ACTIVE METALS AND METAL HYDRIDES

a. Introduction

- Alcohols react with active metals (M = Li, K, Na,...) and metal hydrides (LiH, NaH, KH, etc.) to give H_2 gas and metal alkoxides.

b. The general Reaction for metals:

$$2ROH + 2M \longrightarrow 2RO^- M^+ + H_2$$

Ex:

$$2CH_3OH + 2Li \longrightarrow 2CH_3O^-Li^+ + H_2$$

- The product is Lithium methoxide.

c. The general Reaction for metal hydrides:

$$ROH + MH \longrightarrow RO^-M^+ + H_2$$

Ex:

$$CH_3CH_2OH + KH \longrightarrow CH_3CH_2O^-K^+ + H_2$$

- The product is potassium ethoxide.

- **Overall Summary on Elimination and Substitution Reactions of Alcohols:**

- **Elimination Reactions: H₂SO₄, 85% H₃PO₄, POCl₃, TsOH.**

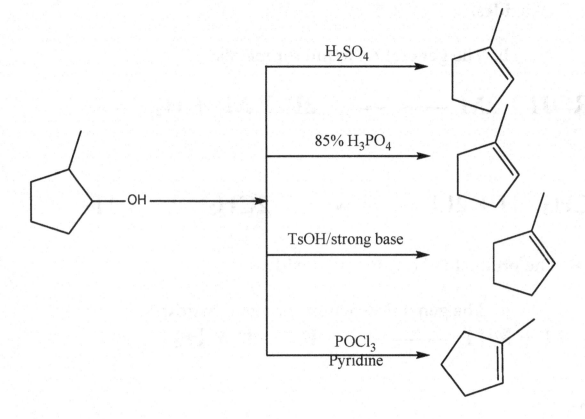

- **Substitution Reactions: HX, SOCl₂, PBr₃, PbCl₃, TsCl.**

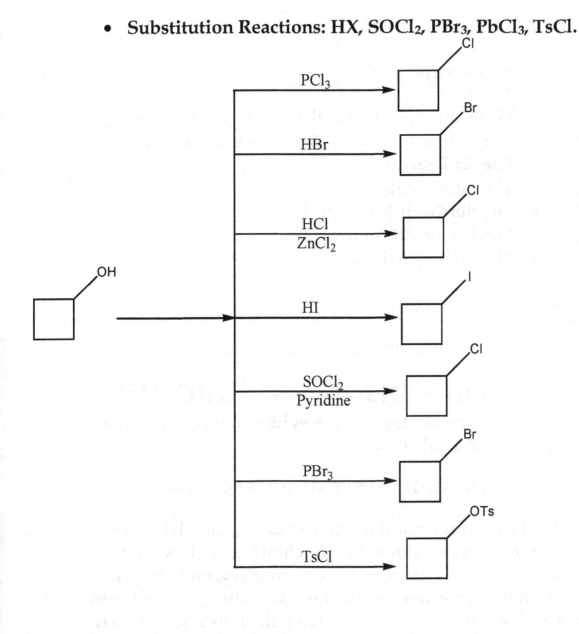

- See Fig. _____, page _____. Do problems on page_____.

F. TESTING FOR ALCOHOLS

1. INTRODUCTION

- There are several ways that can be used to detect the presence of an alcohol. We will focus on the following:
- **The Na Reaction.**
- **The Lucas Test.**
- **The Bordwell-Wellman Test.**
- **The Esterification Reaction.**
- **The $FeCl_3$ reaction.**

3. THE Na REACTION

- **The General Reaction is:**

$$2ROH + 2Na \longrightarrow 2RO^- Na^+ + H_2$$

- **The evolution of H_2 gas is the evidence supporting the presence of an alcohol.**

3. THE LUCAS TEST: A SUBSTITUTION REACTION

- The **Lucas reagent** is a solution made by dissolving **zinc chloride in concentrated hydrochloric acid.** It is used to differentiate between **primary, secondary, and tertiary alcohols.** While **tertiary alcohols** are **swiftly** converted **to alkyl halides** in the reaction, **secondary alcohols** take from **4 to 5 minutes** to react. For **primary alcohols, several hours** are needed.

- Please, see Section F. 5. b.

- **The general reaction is:**

$$ROH + H\!-\!Cl \xrightarrow{\ ZnCl_2\ } RCl + H_2O$$

- **A positive test is based on the rate of the reaction.**

4. THE BORDWELL-WELLMAN TEST: AN OXIDATION REACTION

- This test is used to detect the presence of **primary and secondary** alcohols. The reagent (**chromic acid, H₂CrO₄**) is a solution of chromic anhydride, CrO_3 in sulfuric acid, a strong oxidizing agent. In this reaction, the alcohol is oxidized by **Cr(VI)** which is reduced to **green Cr³⁺**. **Tertiary alcohols** are not oxidized by this reagent.

- **The general reaction is:**

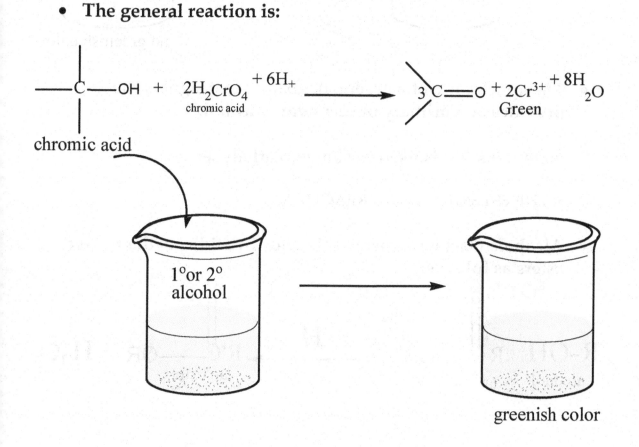

greenish color

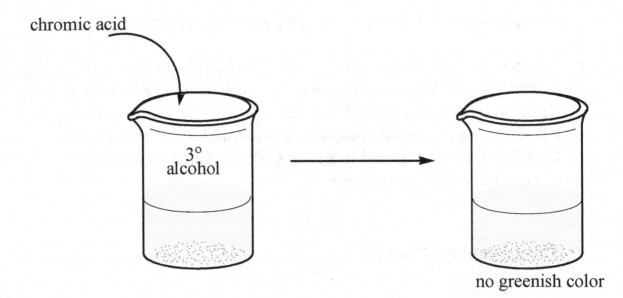

chromic acid

3°
alcohol

no greenish color

- The appearance of a greenish, opaque cast signals the presence of a primary or secondary alcohol.

- Note: This test is used in some breathalyzers.

5. THE ESTERIFICATION REACTION

- Alcohols react with carboxylic acids in acidic media to give esters as follows:

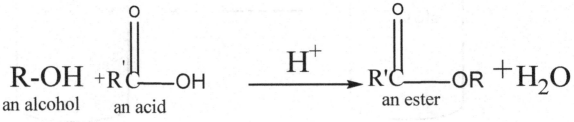

R-OH + an alcohol an acid $\xrightarrow{H^+}$ an ester $+ H_2O$

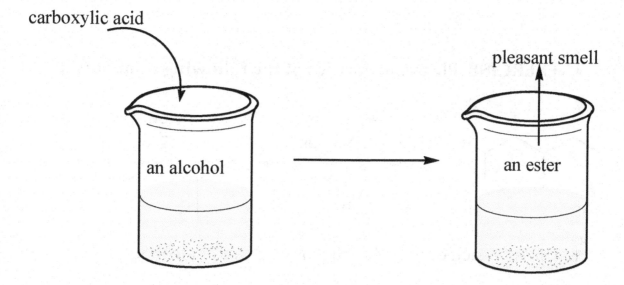

carboxylic acid

an alcohol

pleasant smell

an ester

- The pleasant odor coming from the ester product is evidence that an alcohol has reacted with the acid.

6. THE REACTION OF PHENOLS AND ENOLS WITH FeCl₃

- This test is used to differentiate between phenols/enols from alcohols. Indeed phenols and enols react with FeCl₃ to give various iron complexes (pink, violet, or green). Ordinary alcohols do not react with this reagent. The general reaction is

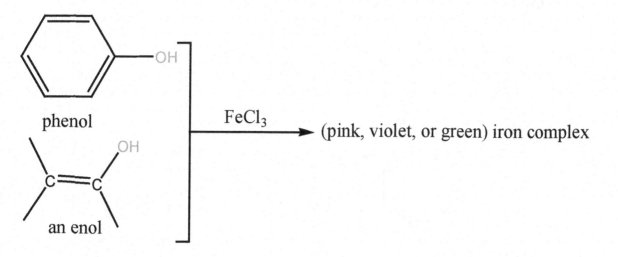

phenol

OH

an enol

$FeCl_3$ → (pink, violet, or green) iron complex

- A positive test results in the appearance of a colorful complex.

- See Key Concepts, pages _____ - _____.

- **EXERCISE :Please, name each of the following compounds:**

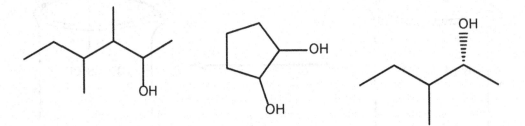

1. Classify each of the following as 1°, 2°, or 3° alcohol.

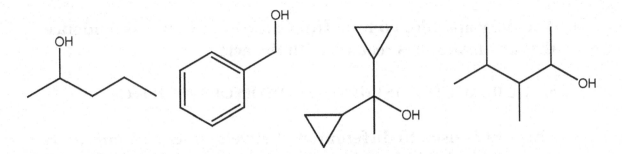

2. For each of the following reactions, give the structure or formula of the missing piece (product(s), catalyst(s), or reactant (s)):

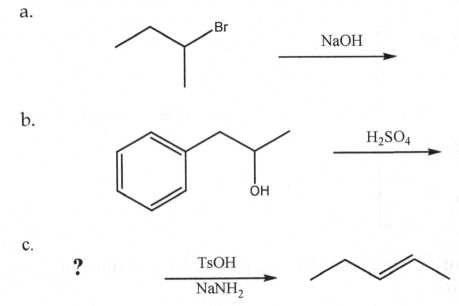

a.

Br   NaOH

b.

                     H$_2$SO$_4$

OH

c.

?   TsOH
      ⟶
      NaNH$_2$

d.

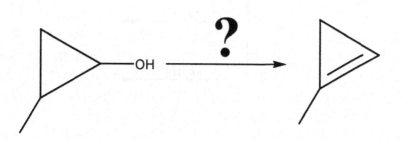

? $\xrightarrow{\text{85\% H}_3\text{PO}_4}$

e.

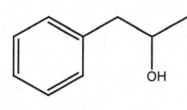

?

f.

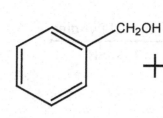

$\xrightarrow{\text{HCl/ZnCl}_2}$

g.

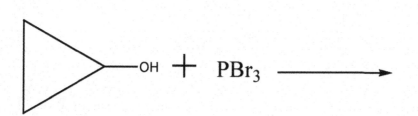

+ SOCl$_2$ $\xrightarrow{\text{Pyridine}}$

h.

CH$_2$OH

—OH + PBr$_3$ $\longrightarrow$

315

i.

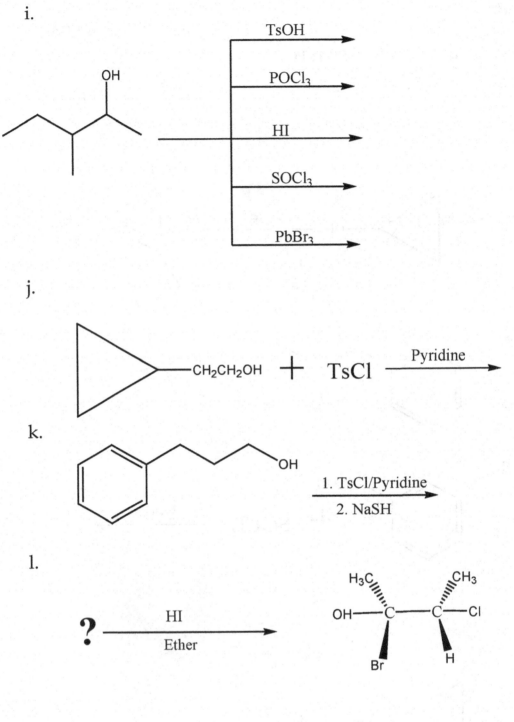

j.

k.

l.

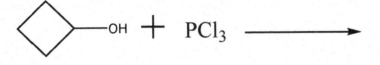

m.

n.

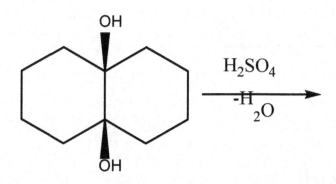

H_2SO_4

$-H_2O$

o.

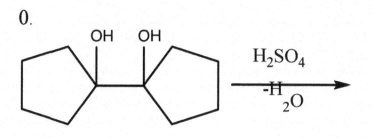

H_2SO_4

$-H_2O$

p.

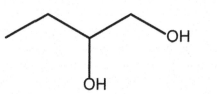

H_2SO_4

$-H_2O$

q.

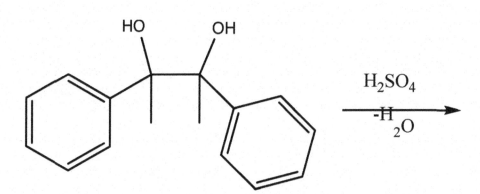

H_2SO_4

$-H_2O$

r.

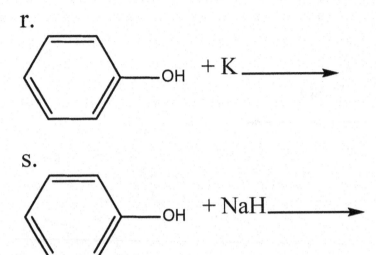

s.

OCHEM I UNIT 13: ETHERS, EPOXIDES, PHENOLS, AND THIOLS

A. GENERAL STRUCTURE OF ETHERS

1. THE ETHER FUNCTIONAL GROUP

- Organic substances that contain **R—O—R′** are called **ethers**.

Ex: $CH_3CH_2OCH_3$; $CH_3CH_2OCH_2CH_3$

2. TYPES OF ETHERS

   a. **Symmetrical Ethers: R = R′ in ROR′**

Ex: $CH_3CH_2OCH_2CH_3$

   b. **Unsymmetrical Ethers: R ≠ R′ in ROR′**

Ex: $CH_3CH_2OCH_3$

B. NAMING ETHERS

1. IUPAC NAMES
- According to IUPAC rules, the group **RO—** is called the **alkoxy group.** Therefore, ethers are named as **alkoxyalkanes.**

Ex:

CH_3OCH_3 = methoxymethane

$CH_3CH_2OCH_2CH_3$ = ethoxyethane

$CH_3CH_2OCH_3$ = methoxyethane

- See examples and problems on pages _____ - _____ .

2. NAMES FOR CYCLIC ETHERS

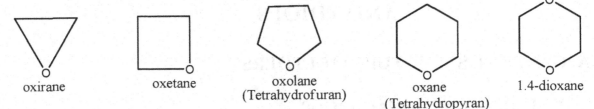

| oxirane | oxetane | oxolane (Tetrahydrofuran) | oxane (Tetrahydropyran) | 1.4-dioxane |

3. COMMON NAMES

- Ethers can be also named as **alkyl ethers.** Alkyl groups are listed **alphabetically.**

Ex:

CH_3OCH_3 = **dimethyl ether**

$CH_3CH_2OCH_2CH_3$ = **Diethyl ether**

$CH_3CH_2OCH_3$ = **Ethyl methyl ether**

C. PHYSICAL PROPERTIES OF ETHERS

1. BONDING IN ETHERS

- Ethers are **bent, polar** molecules **like water.**

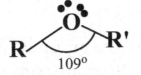

2. BOILING AND MELTING POINTS OF ETHERS

- Ethers are polar substances. Therefore they have **dipole-dipole interactions and van der Waals forces** between their molecules. However, they cannot have H bonding like alcohols and water since they do not contain –OH groups. As a result, they have **lower** BP and MP than alcohols of comparable molar masses. But, they have **higher** BP and MP than normal alkanes.

Ex: increasing BP of some compounds.

$$CH_3CH_2CH_2CH_3 < CH_3OCH_2CH_3 < CH_3CH_2CH_2CH_2OH$$

3. SOLUBILITIES OF ETHERS

- The solubilities of ethers are similar to those of alcohols:
  - **Ethers with 5 carbons or less are soluble in water.**
  - **Ethers with more than 5 carbons are insoluble in water.**
  - **As organic substances, all ethers are soluble in organic solvents.**

- See Table _____, page _____.

- Do Problem _____ page _____.

D. IMPORTANT ETHERS

- **Diethyl ether or ethoxyethane: $CH_3CH_2OCH_2CH_3$**

- **Crown ethers:**

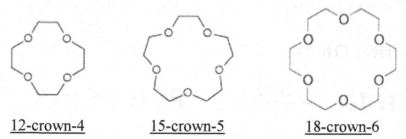

12-crown-4   15-crown-5   18-crown-6

- **Question: How do you name crown ethers?**

- **Tetrahydrofuran (THF): A good polar aprotic solvent in SN2 Reactions.**

tetrahydrofuran

- **1,4-dioxane:** Another polar aprotic solvent.

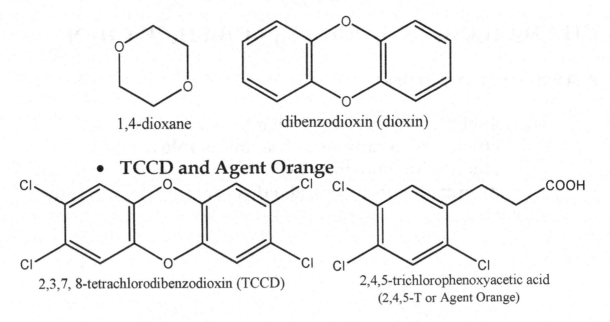

1,4-dioxane   dibenzodioxin (dioxin)

- **TCCD and Agent Orange**

2,3,7, 8-tetrachlorodibenzodioxin (TCCD)

2,4,5-trichlorophenoxyacetic acid
(2,4,5-T or Agent Orange)

E. SYNTHESIS OF ETHERS

1. INTRODUCTION

- Ethers can be easily prepared by using **SN2 reactions** with **1°** alkyl halides.

- **Symmetrical Ethers:**

$$RCH_2Cl + {}^-OCH_2R \xrightarrow{\text{SN2}} RCH_2OCH_2R$$

Ex:

$$CH_3CH_2Cl + {}^-OCH_2CH_3 \xrightarrow{\text{SN2}}$$

- **Unsymmetrical Ethers**

$$RCH_2Cl + {}^-OCH_2R' \xrightarrow{\text{SN2}} RCH_2OCH_2R'$$

Ex:

$$CH_3CH_2Cl + {}^-OCH_2CH_2CH_3 \xrightarrow{\text{SN2}}$$

2. THE WILLIAMSON SYNTHESIS OF ETHERS

- This method uses less hindered alkyl halides: 1° halides.
- It proceeds in two steps:

- Step 1: Formation of the nucleophile from an alcohol:

$$ROH + NaH \ (LiH, KH) \longrightarrow RO^- \ Na^+ + H_2$$
or
$$2ROH + 2Na \ (or \ K \ or \ Li) \longrightarrow 2Na^+ \ 2RO^- + H_2$$

- Step 2: SN2 reaction of the RO⁻ nucleophile with a 1° halide.

$$RO^- + R'\text{-}X \xrightarrow{\text{THF}} ROR' + X^-$$

Ex: Please, give two reactants you could use to synthesize the following:

$$? \longrightarrow CH_3CH_2OCH_3$$

3. SYNTHESIS OF SYMMETRICAL ETHERS USING AN ALCOHOL

$$2ROH \xrightarrow[140^{\circ}C]{H_2SO_4} ROR$$

Ex:

$$2CH_3CH_2OH \xrightarrow[140^{\circ}C]{H_2SO_4} CH_3CH_2OCH_2CH_3 + H_2O$$

4. SYNTHESIS OF SYMMETRICAL AND UNSYMMETRICAL ETHERS USING TWO DIFFERENT ALCOHOLS

$$ROH + R'\text{-}OH \xrightarrow[140^{\circ}C]{H_2SO_4} ROR + ROR' + R'OR'$$

Ex:

$$CH_3OH + CH_3CH_2OH \xrightarrow[140^{\circ}C]{H_2SO_4} CH_3OCH_3 + CH_3OCH_2CH_3$$
$$+ CH_3CH_2OCH_2CH_3$$

5. ACID –CATALYZED ADDITION OF AN ALCOHOL TO AN ALKENE

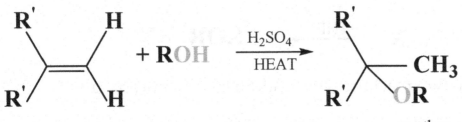

Ex:

an ether

324

- **Reaction Mechanism:**

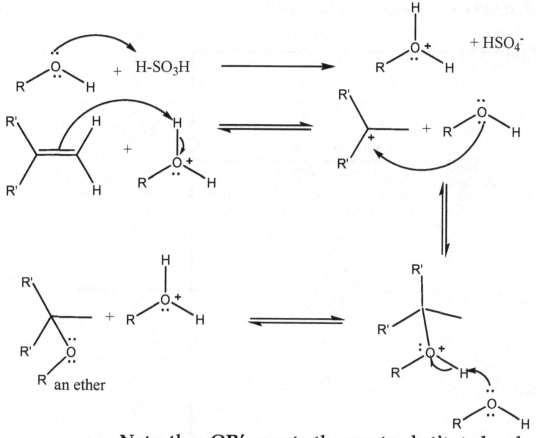

- **Note: the –OR′ goes to the most substituted carbon because the carbocation intermediate formed at that carbon is the most stable.**

6. AN OTHER METHOD USING A SILVER OXIDE CATALYST

$$ROH \xrightarrow[Ag_2O]{R'X} ROR' + AgX$$

Ex: Which alcohol would you use to synthesize the following compound?

$$? \xrightarrow[Ag_2O]{CH_3CH_2Br} CH_3OCH_2CH_3 + AgBr$$

- Read pages _____ - _____.

7. SYNTHESIS OF ETHERS: A SUMMARY

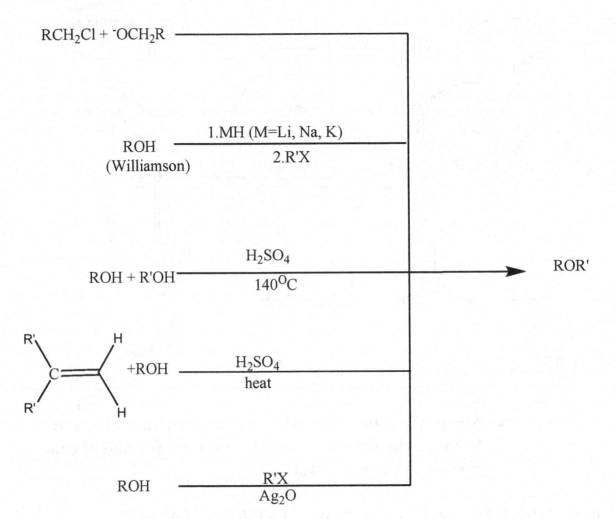

F. REACTIONS OF ETHERS

1. INTRODUCTION

- **Recall the general structure of ethers:**

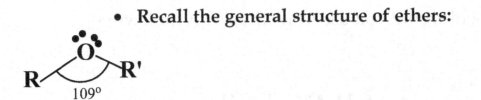

- The —**OR** group is a poor leaving group. As a result, ethers are very stable (unreactive). They do not undergo direct SN or E reactions. This is why they are used as solvents. However, they do undergo cleavage reactions in the presence of strong acids HX (X = Br or I). Ethers are highly volatile and flammable. They can cause severe frostbite. "Aged" ethers are also dangerous, because they can explode. Please, see auto oxidation of Ethers in Section A.3

2. ACIDIC CLEAVAGE OF ETHERS: HX (X = Br OR I)

a. General Reaction

- This cleavage reaction occurs with **primary (SN2), secondary, and tertiary (SN1) groups** on the ether oxygen.

$$ROR' + 2H\!-\!X \longrightarrow R\!-\!X + R'\!-\!X + H_2O$$

Ex:

$$CH_3CH_2OCH_3 + 2H\text{-}I \longrightarrow CH_3CH_2I + CH_3I + H_2O$$

    b. See reaction mechanism on page _____. The 2 –OR bonds are cleaved in 2 separate steps.

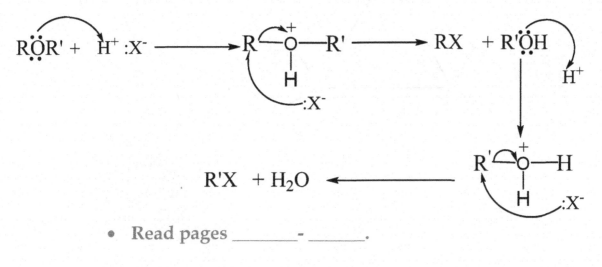

- Read pages _____ - _____ .

- Do problems on page _____ .

3. AUTOOXIDATION OF ETHERS EXPOSED TO ATMOSPHERIC O₂.

- Ethers exposed to atmospheric oxygen undergo **spontaneous oxidation** as follows and can explode violently to give a hydroxyperoxide and a dialkyl peroxide.

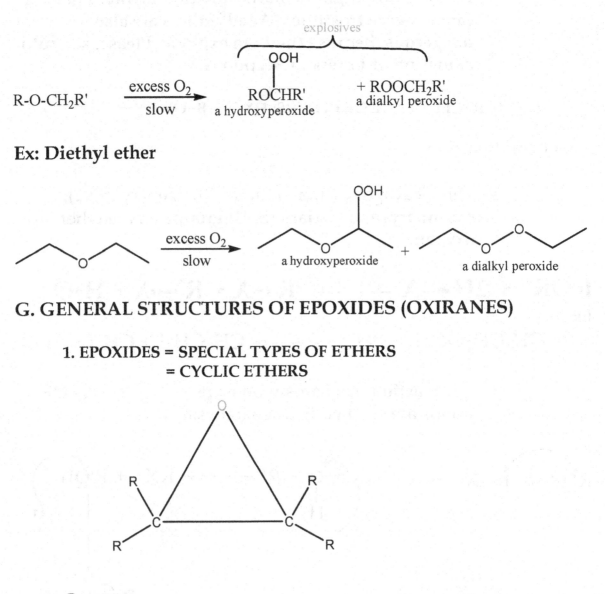

Ex: Diethyl ether

G. GENERAL STRUCTURES OF EPOXIDES (OXIRANES)

1. EPOXIDES = SPECIAL TYPES OF ETHERS
    = CYCLIC ETHERS

- See page _____.

2. OXIRANES

- Epoxides are also known as **oxiranes.**

H. NAMING EPOXIDES

1. IUPAC NAMES

a. Monosubstituted Epoxides

- Monosubstituted epoxides are named as **derivatives of oxirane. Do not number the carbons.**

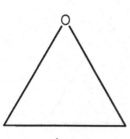

oxirane

Ex:

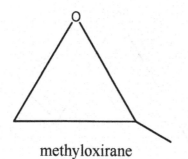

methyloxirane

b. Disubstituted Epoxides

- The O and the 2 carbons are numbered as follows:

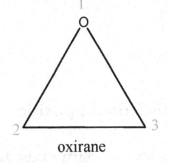

oxirane

Ex:

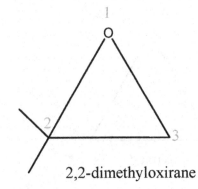

2,2-dimethyloxirane

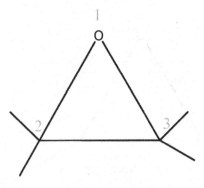

2,2,3,3-tetramethyloxirane

c. Epoxy Alkanes and Cycloalkanes

- Epoxides can be also named as **epoxy alkanes.** In this case, the oxygen is **not assigned** a number. Here are some examples:

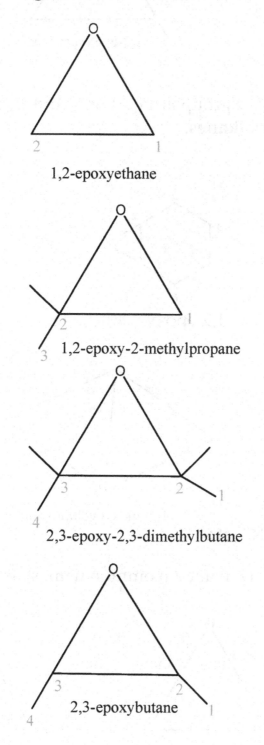

1,2-epoxyethane

1,2-epoxy-2-methylpropane

2,3-epoxy-2,3-dimethylbutane

2,3-epoxybutane

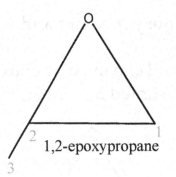

1,2-epoxypropane

- When the epoxy group is a **part of a ring,** epoxides are named as **epoxycycloalkanes.**

Ex:

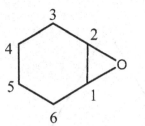

1,2-epoxycylohexane

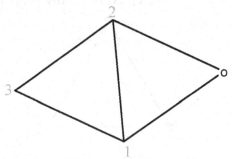

1,2-epoxycyclopropane

2. COMMON NAMES

- Epoxides can be named (common names) as **alkene oxides.**

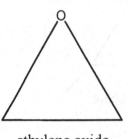

ethylene oxide

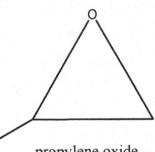

propylene oxide

I. PHYSICAL PROPERTIES OF EPOXIDES
1. BONDING IN EPOXIDES

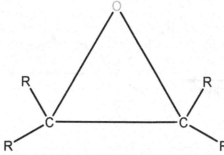

- Epoxides are polar substances like ethers. So they have similar properties as ethers: MP, BP, Solubility.

J. IMPORTANT EPOXIDES

- See and read page _____ about Eplerenone (cardiovascular drug) and Tiotropium bromide (antismoke drug).

J. SYNTHESIS OF EPOXIDES

1. USING A HALOHYDRIN

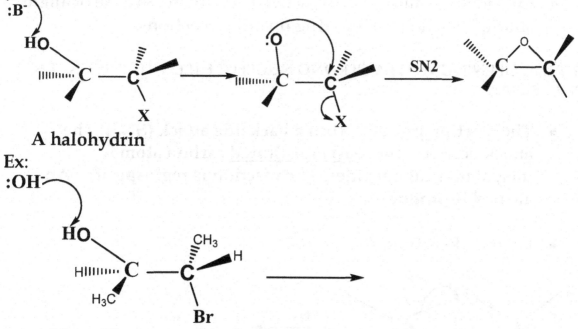

A halohydrin

Ex:

- Note: The stereochemistry of the reactant is conserved.

- Do problem on page _____.

2. USING O_2/Ag_2O CATALYST

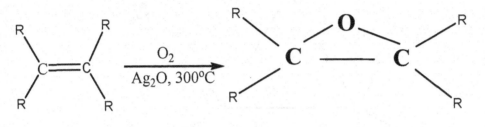

Ex:

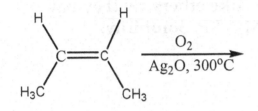

K. REACTIONS OF EPOXIDES

1. INTRODUCTION

- Epoxides are much more reactive than ethers because of **angle strain**. They do undergo **ring opening reactions**.

2. RING OPENING WITH A STRONG NUCLEOPHILE: OH^-, OR^-, CN^-, RS^-, NH_3, $RC{\equiv}C{:}^-$.

- **The reaction proceeds thru a backside attack (SN2). The attack occurs at the <u>least substituted</u> carbon atom for unsymmetrical epoxides. The reaction is regiospecific. An alcohol is produced.**

- **General Reaction:**

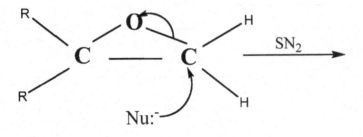

Ex:

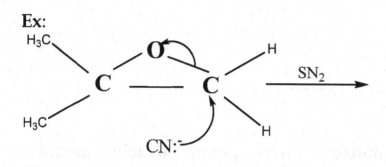

3. RING OPENING WITH A STRONG ACID HZ: HCl, HBr, HI

a. Using Strong acid HZ (HCl, HBr, HI)

- Ring opening of epoxides with strong acids is also regiospecific. In this case, the attack occurs at <u>the most substituted</u> carbon for unsymmetrical epoxides. The reaction can be either SN1 or SN2.

- **General Reaction:**

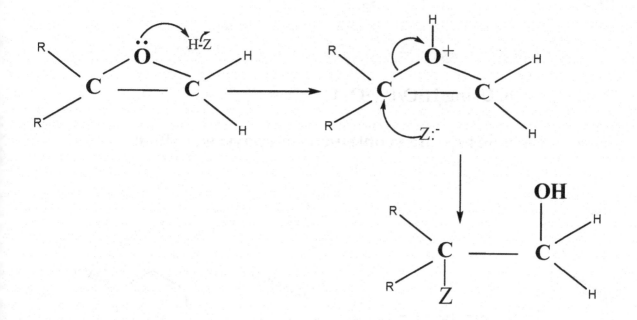

Ex:

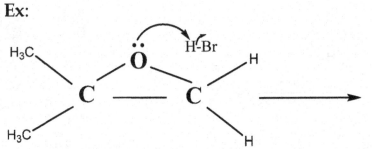

- **Note: Similarly, alcohols react with epoxides in acidic media as follows:**

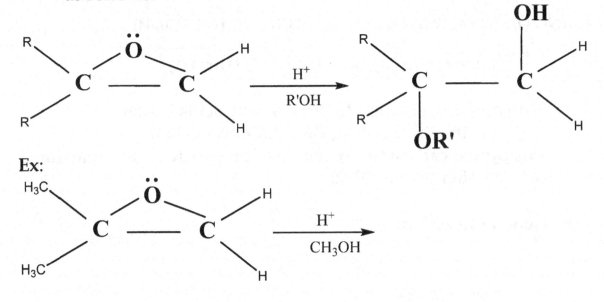

Ex:

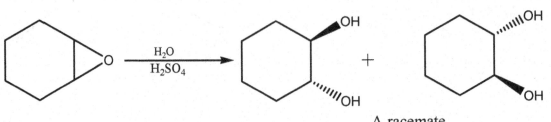

- See pages _____ – _____. Do all associated problems.

b. Using H_2O/H_2SO_4 Catalyst

- **A racemic mixture is obtained for epoxycycloalkanes.**

Ex:

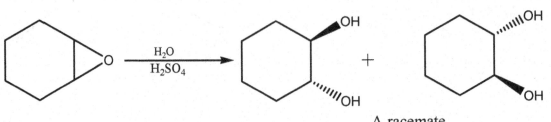

A racemate

- **Note: If an achiral substrate is used, then the product is <u>always</u> achiral.**

4. ARENE OXIDE RING OPENING

a. Introduction

- **In general, arene oxides undergo <u>two types</u> of reactions:**
- **Addition reactions with nucleophiles in acidic media to give a pair of enantiomers**
- **Internal rearrangement reactions in acidic media to give phenols.**

b. Addition Reactions

- **The general reaction: 2 enantiomers are obtained**

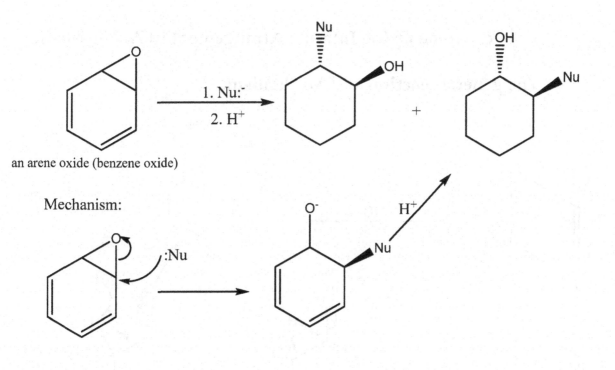

an arene oxide (benzene oxide)

Mechanism:

Ex:

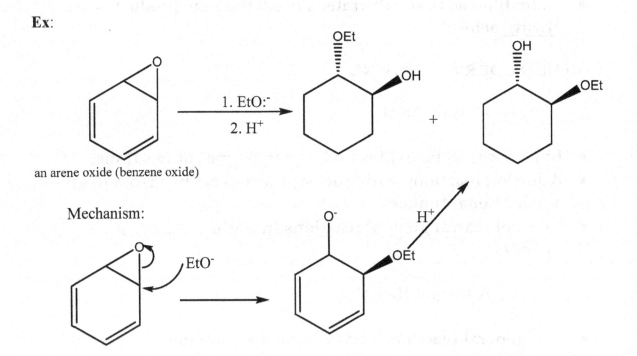

an arene oxide (benzene oxide)

Mechanism:

c. Arene Oxide Internal Arrangement in Acidic Media

- ## The general reaction and Mechanism:

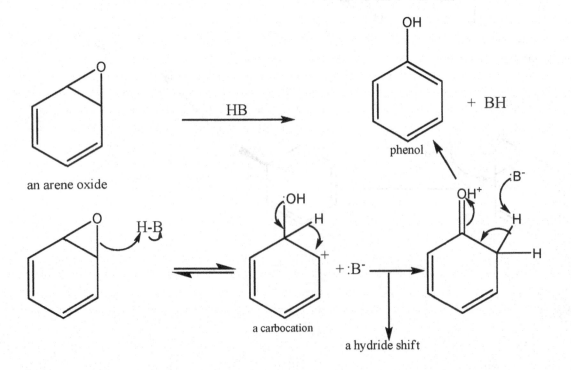

an arene oxide

Ex:

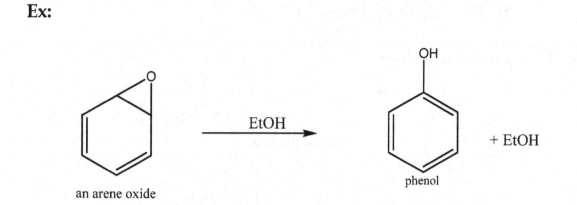

an arene oxide

phenol

+ EtOH

L. GENERAL STRUCTURES OF PHENOLS

1. GENERAL STRUCTURE

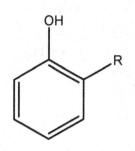

2. PREPARATION OF PHENOL

a. Phenols from Cumene

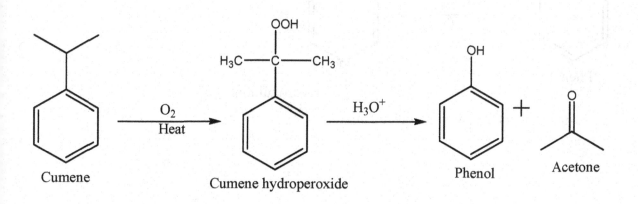

Cumene

Cumene hydroperoxide

Phenol

Acetone

b. Phenols from Phenol Ethers

- ## The general reaction:

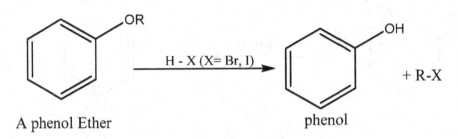

A phenol Ether   phenol

Ex:

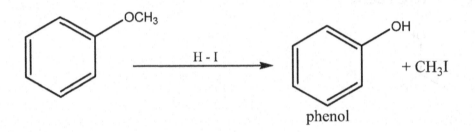

3. OXIDATION OF PHENOL

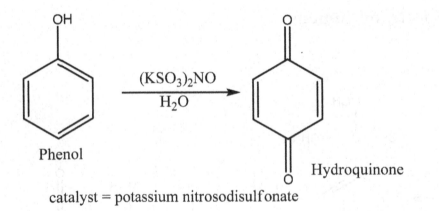

Phenol   Hydroquinone

catalyst = potassium nitrosodisulfonate

4. SOME IMPORTANT PHENOLS

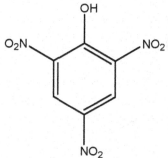

2,4,6-Trinitrophenol (TNP)
(picric acid = explosive)

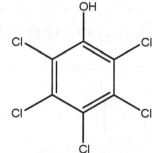

Pentachlorophenol
(wood preservative)

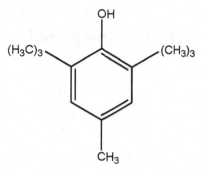

Butylated hydroxytoluene (BHT)
(food preservative)

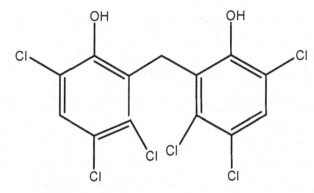

Hexachlorophene = an antiseptic

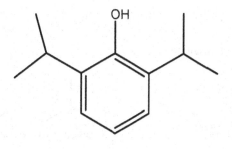

Propofol = an anesthetic

M. GENERAL STRUCTURES OF THIOLS AND SULFIDES

1. GENERAL STRUCTURE

$$R - SH = \text{Thiol or mercaptan}$$

$$R - S - R' = \text{Sulfide}$$

- **Note: R-SH group = mercapto group**

2. NOMENCLATURE

- Thiols are named as **alkanethiols.**

Ex: **CH₃SH is methanethiol (methyl mercaptan)**

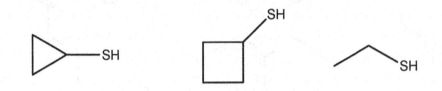

- Simple sulfides are named as sulfides. Complex sulfides are named as **alkylthio.**

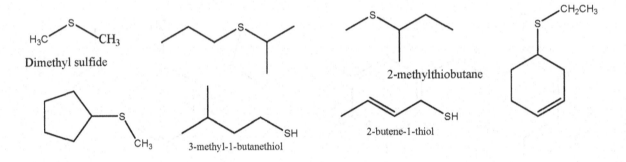

Dimethyl sulfide

2-methylthiobutane

3-methyl-1-butanethiol

2-butene-1-thiol

3. SYNTHESIS OF THIOLS

a. Using Hydrogen Sulfide: The General Reaction:

$$R\text{-}X + Na^+ SH^- \longrightarrow R\text{-}SH + NaX$$

Ex:

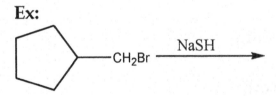

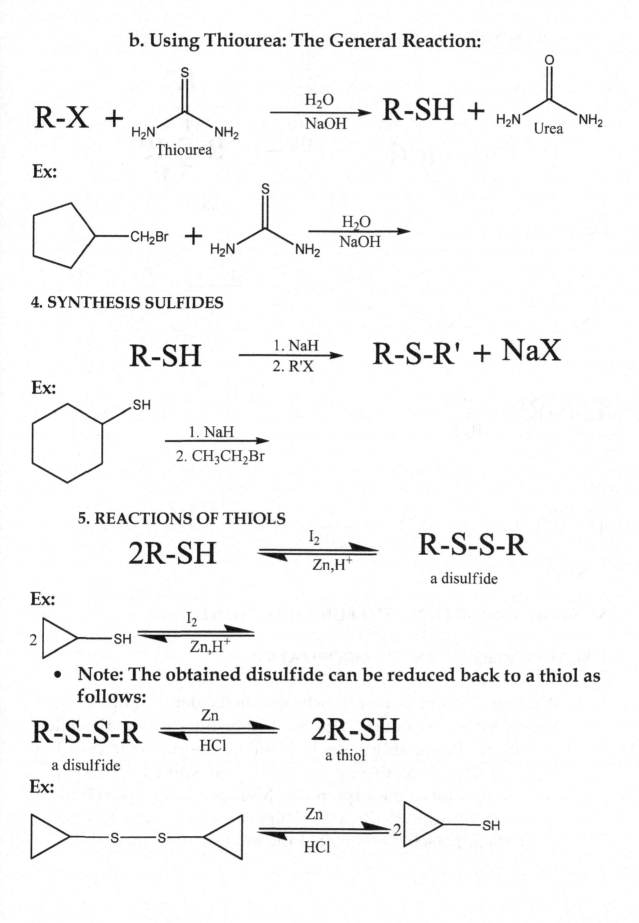

b. Using Thiourea: The General Reaction:

$$R\text{-}X + \underset{\underset{\text{Thiourea}}{H_2N}\quad NH_2}{\overset{S}{\|}} \xrightarrow[\text{NaOH}]{H_2O} R\text{-}SH + \underset{\underset{\text{Urea}}{H_2N}\quad NH_2}{\overset{O}{\|}}$$

Ex:

$$\text{cyclopentyl-}CH_2Br + \underset{H_2N\quad NH_2}{\overset{S}{\|}} \xrightarrow[\text{NaOH}]{H_2O}$$

4. SYNTHESIS SULFIDES

$$R\text{-}SH \xrightarrow[\text{2. R'X}]{\text{1. NaH}} R\text{-}S\text{-}R' + NaX$$

Ex:

cyclohexyl-SH $\xrightarrow[\text{2. }CH_3CH_2Br]{\text{1. NaH}}$

5. REACTIONS OF THIOLS

$$2R\text{-}SH \underset{Zn,H^+}{\overset{I_2}{\rightleftharpoons}} \underset{\text{a disulfide}}{R\text{-}S\text{-}S\text{-}R}$$

Ex:

$$2 \text{ cyclopropyl-}SH \underset{Zn,H^+}{\overset{I_2}{\rightleftharpoons}}$$

- **Note: The obtained disulfide can be reduced back to a thiol as follows:**

$$\underset{\text{a disulfide}}{R\text{-}S\text{-}S\text{-}R} \underset{\text{HCl}}{\overset{Zn}{\rightleftharpoons}} \underset{\text{a thiol}}{2R\text{-}SH}$$

Ex:

$$\text{cyclopropyl-}S\text{-}S\text{-cyclopropyl} \underset{\text{HCl}}{\overset{Zn}{\rightleftharpoons}} 2 \text{ cyclopropyl-}SH$$

6. REACTIONS OF SULFIDES

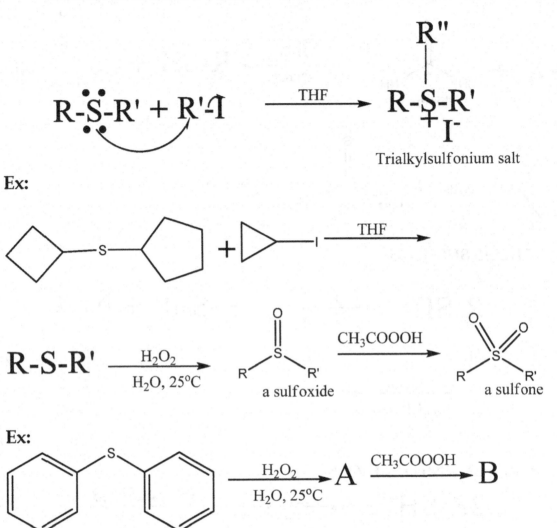

R-S-R' + R'-I $\xrightarrow{\text{THF}}$ R-S-R' I⁻ (R" on sulfur)

Trialkylsulfonium salt

Ex:

[cyclobutyl-S-cyclopentyl] + [cyclopropyl-I] $\xrightarrow{\text{THF}}$

R-S-R' $\xrightarrow[\text{H}_2\text{O, 25°C}]{\text{H}_2\text{O}_2}$ a sulfoxide $\xrightarrow{\text{CH}_3\text{COOOH}}$ a sulfone

Ex:

[phenyl-S-phenyl] $\xrightarrow[\text{H}_2\text{O, 25°C}]{\text{H}_2\text{O}_2}$ A $\xrightarrow{\text{CH}_3\text{COOOH}}$ B

M. SOME IMPORTANT SULFUR COMPOUNDS

1. METHYL MERCAPTAN GAS ODORIZATION

- Whereas thiols have bad stench, low molecular weight hydrocarbons are **odorless** gases. As a result, a leak of the latter dangerous, **flammable** gases in a storage is virtually impossible to detect. Therefore, these gases are usually **spiked** with a small amount of **thiol or mercaptan (Ex. Methanethiol, CH₃SH)** to give them a strong foul odor… Thus, one can detect a gas leak when it happens by the smell of the mixture. If needed, the mixture can be deodorized by passing it through **sodium methoxide (CH₃ONa)** in an acid-base reaction:

$$CH_3-SH \ + \ CH_3ONa \longrightarrow CH_3-SNa \ + \ CH_3OH$$

methanethiol

- **Methyl mercaptan** is also the precursor of pesticides, jet fuels, plastics, amino acids and methionine syntheses.

2. SULFUR COMPOUNDS IN SKUNK SPRAY

- Skunk spray consists of a mixture of several volatile stenchy thiols. The **3 major** compounds responsible for skunk spray odor are:

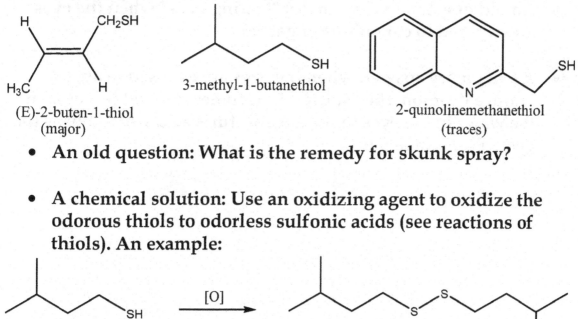

(E)-2-buten-1-thiol
(major)

3-methyl-1-butanethiol

2-quinolinemethanethiol
(traces)

- **An old question: What is the remedy for skunk spray?**

- **A chemical solution: Use an oxidizing agent to oxidize the odorous thiols to odorless sulfonic acids (see reactions of thiols). An example:**

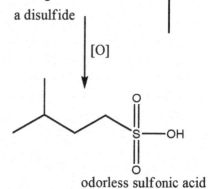

odorous 3-methyl-1-butanethiol   a disulfide

odorless sulfonic acid

3. SULFUR COMPOUNDS IN ONIONS AND GARLIC

- The major sulfur compound found in onions and garlic is **alliin**. When garlic is "injured", an enzyme from the inside of the garlic called **alliinase** comes out to the rescue and converts alliin to **allicin** according to the following reaction:

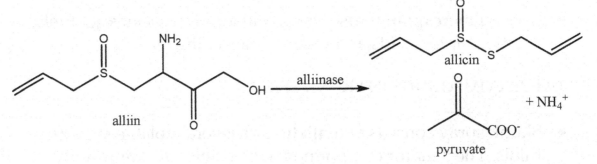

- **An old question: What causes "tearing eyes" when the eyes are exposed to cut onions or garlic?**

- **A Chemistry answer: when the eyes are exposed to an "injured" onion, the "sulfur" in it comes out and reacts with the water in the eyes to produce sulfuric acid which burns the eyes. Tears come out.**

- See Key Concepts, pages _____ – _____.

• EXERCISE

1. Please, name each of the following compounds:

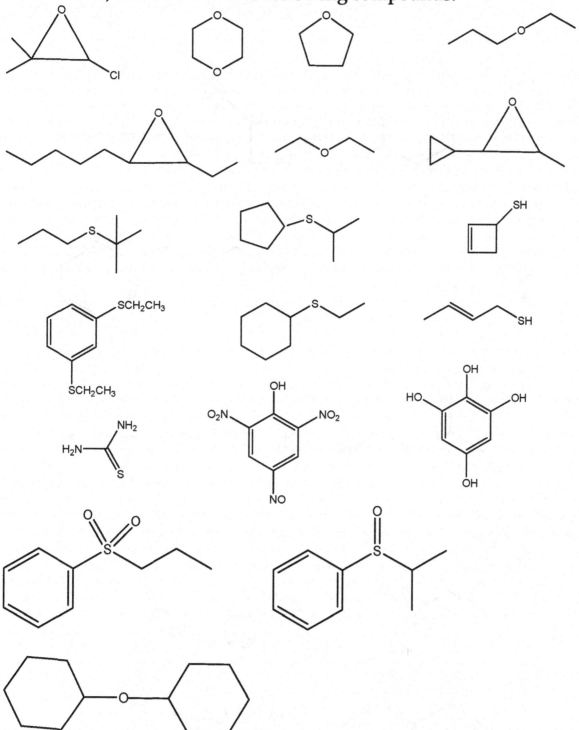

2. For each of the following reactions, give the structure or formula of the missing piece (product(s), catalyst(s), or reactant (s):

a.

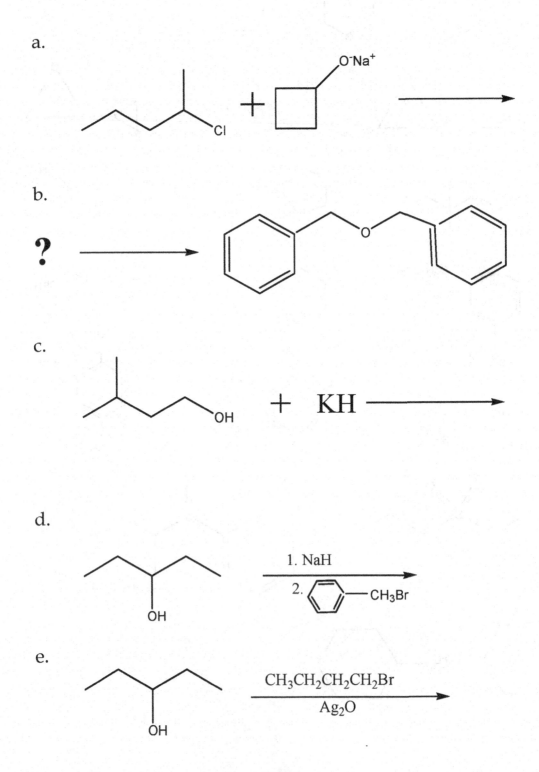

b.

c.

d.

1. NaH

2. ⬡—CH₃Br

e.

CH₃CH₂CH₂CH₂Br

Ag₂O

f.

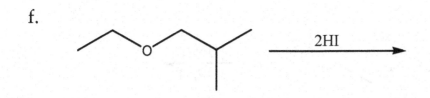

g.

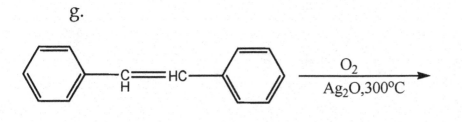

h.

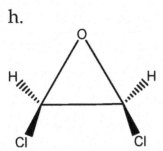

i.

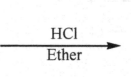

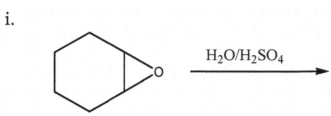

j.

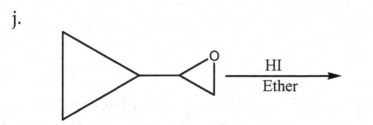

k.

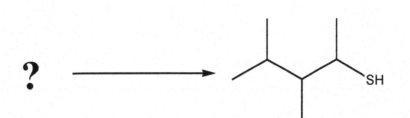

l.

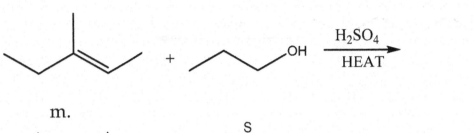

$\xrightarrow{\text{H}_2\text{SO}_4}$
HEAT

m.

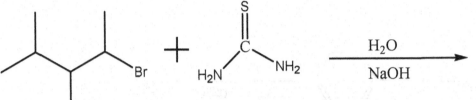

$\xrightarrow[\text{NaOH}]{\text{H}_2\text{O}}$

n.

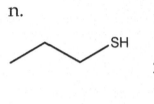

1. NaH

2. 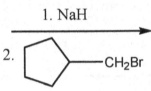 —CH$_2$Br

o.

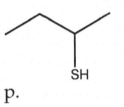

$\xrightarrow[\text{Zn, H}^+]{\text{I}_2}$

p.

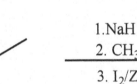

1. NaH
2. CH$_3$Br
3. I$_2$/Zn, H$^+$

q.

SH $\xrightarrow[\text{H}_2\text{O, 25}^o\text{C}]{\text{H}_2\text{O}_2}$ **A** $\xrightarrow{\text{CH}_3\text{COOOH}}$ **B**

r.

350

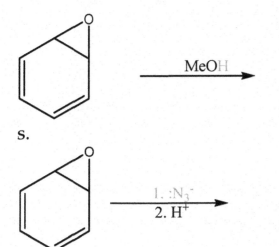

s.

$\xrightarrow{\text{MeOH}}$

t.

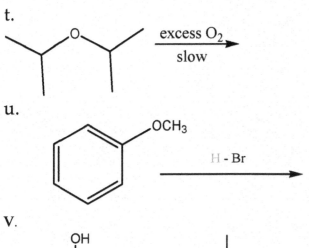

$\xrightarrow[\text{2. H}^+]{\text{1. :N}_3^-}$

u.

$\xrightarrow[\text{slow}]{\text{excess O}_2}$

v.

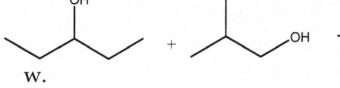

OCH$_3$

$\xrightarrow{\text{H - Br}}$

w.

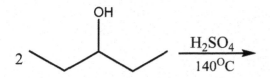

OH

$+$

OH

$\xrightarrow[\text{140}^\text{O}\text{C}]{\text{H}_2\text{SO}_4}$

2

OH

$\xrightarrow[\text{140}^\text{O}\text{C}]{\text{H}_2\text{SO}_4}$

OCHEM I UNIT 14: ALKENES

A. INTRODUCTION

1. DEFINITION

- **Alkenes** belong to the **second** class of hydrocarbons.
- **Alkenes** are **unsaturated hydrocarbons** that contain at least one **C-C double bond.**
- **Alkenes** are also called **olefins**.

2. KINDS OF ALKENES

- There are **two kinds** of alkenes:

    o **Straight chain alkenes: acyclic**
    o **Cycloalkenes: cyclic**

3. GENERAL MOLECULAR FORMULA OF ALKENES

$$C_nH_{2n}$$

n = Total number of carbon atoms in molecule.

- **Note: Have a homologous series.**

Ex: n = 1

      n = 2

      n = 6

4. GENERAL MOLECULAR FORMULA OF CYCLOALKENES: CYCLIC ALKENES

$$C_nH_{2n-2}$$

Ex: C_6H_{10}

cyclohexene

- For cycloalkenes with less than **8 carbons**, only **cis isomers** have been isolated because the trans isomers are too **strained** to exist. However, for higher cycloalkenes, both isomers exist.

5. TYPES OF ALKENES: 4 types

a. Terminal Alkenes

1-butene

b. Internal Alkenes

2-butene

c. Symmetrical Alkenes

3-hexene

d. Unsymmetrical Alkenes

2-pentene

6. PROPERTIES OF THE C-C DOUBLE BOND

- **Recall:**

- The **-C=C-** is a very rigid bond. Therefore, **it cannot be rotated** like a single bond. This gives rise to 2 stereoisomers: *cis and trans.*

Ex:

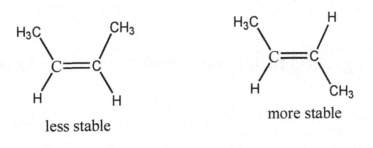

less stable   more stable

- **Do Problem** _____ , **page** _____.

- **Note: The *cis* isomer is less stable than the *trans*. Why?**

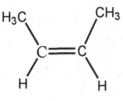

less stable

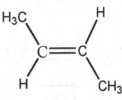

more stable

- **See Table** _____ , **page** _____.

- **Read pages** _____ - _____.

B. DEGREES OF UNSATURATION (DOU)

1. DEFINITION

- The **degree of unsaturation** of an **unsaturated** compound or a **cyclic** compound is the number of **rings and/or multiple bonds (double and triple) in the compound.**

Ex:

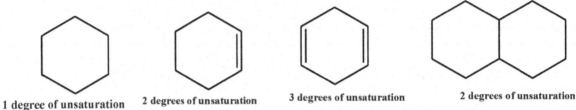

| 1 degree of unsaturation | 2 degrees of unsaturation | 3 degrees of unsaturation | 2 degrees of unsaturation |

- **Recall: The general formula of an alkane is: C_nH_{2n+2}.**

Ex: C_8H_{18}

2. CALCULATING DEGREES OF UNSATURATION

$$DOU = \frac{\text{\# of missing hydrogens in corresponding alkane}}{2}$$

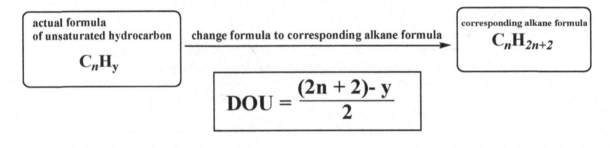

actual formula
of unsaturated hydrocarbon

C_nH_y

change formula to corresponding alkane formula

corresponding alkane formula

C_nH_{2n+2}

$$DOU = \frac{(2n + 2)- y}{2}$$

Ex: C_5H_{10}

3. SOME POSSIBILITIES

#missing Hs	DOU	possibilities
2	1	-1 ring -1 double bond
4	2	-2 rings -2 double bonds -1 triple bond -1 ring and 1 double bond
6	3	-3 rings -3 double bonds -1 triple bond and 1 double bond, etc.

4. DOU FOR COMPOUNDS CONTAINING HALOGENS

- **Add the number of halogens to the number of hydrogens.**

Ex 1: Calculate the DOU of $C_5H_8Cl_2$.

Ex 2: Calculate the DOU of $C_7H_8Br_4$.

5. DOU FOR COMPOUNDS CONTAINING C, H, AND O

- **Ignore the number of oxygens.**

Ex: Calculate the DOU of C_6H_8O.

6. DOU FOR COMPOUNDS CONTAINING C, H, AND N

- **Subtract the number of nitrogens from the number of hydrogens.**

Ex: Calculate the DOU of C_5H_9N.

7. A SUMMARY ON DOU FOR COMPOUNDS CONTAINING HETEROATOMS

- Add the number of halogens to the number of hydrogens.
- Ignore the number of oxygens.
- Subtract the number of nitrogens from the number of hydrogens.

- Read pages _____ – _____.

- Do all problems on page_____.

C. NOMENCLATURE OF THE ALKENES

1. IUPAC NAMES FOR UNBRANCHED ALKENES

- Names end in **– ene** as in **prefix-ene**.
- Prefixes are the same as in naming the alkanes.

n	molecular formula	structural formula	name
2	C_2H_4	$CH_2=CH_2$	Eth-1-**ene**
3	C_3H_6	$CH_3CH=CH_2$	Prop-1-**ene**
4	C_4H_8	$CH_3CH_2CH=CH_2$	But-1-**ene**
5	C_5H_{10}	$CH_3CH_2CH_2CH=CH_2$	Pent-1-**ene**

- Note: Double bonds in structural formulas of alkenes are always written out (shown).

2. ISOMERISM IN ALKENES

n = 4 $CH_3CH_2CH=CH_2$ 1-butene or but-1-ene

$CH_3CH=CHCH_3$ 2-butene or but-2-ene

3. IUPAC RULES FOR NAMING BRANCHED AND UNBRANCHED ALKENES

a. Select **the longest continuous carbon chain (parent or main chain)** that contains **both carbons of the double bond.**

Ex:

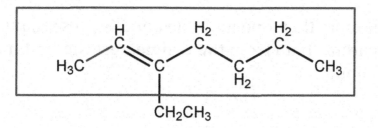

b. Number the carbons of the parent chain from **the end nearest the double bond** so that the carbon atoms in that bond have the **lowest possible** numbers.

Ex:

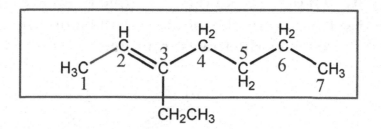

c. If the double bond is **equidistant** from both ends of the parent chain, number the carbons from the **end nearest the first branch point.**

Ex:

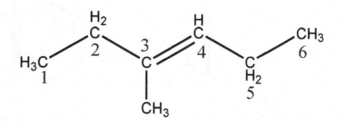

d. Indicate the position of the double bond using the **lowest numbered carbon atom in the double bond.**

Ex:

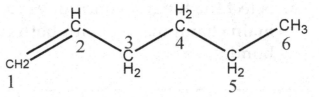

e. If more than one multiple bond is present, number the parent chain **from the end nearest the first multiple bond.**

Ex:

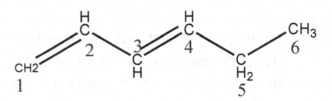

f. If a double bond and a triple bond are equidistant from both ends of the parent chain, the **double bond receives the lowest number.**

Ex:

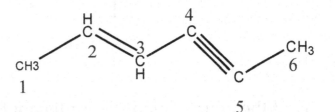

• **Note: For cycloalkenes, the "1" is usually omitted.**

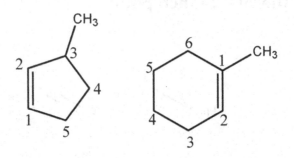

3-methylcyclopentene   1-methylcyclohexene

• **Read pages** _____ - _____.
• **Do problems on pages** _____.

360

4. IUPAC RULES FOR NAMING DIASTEREOMERS

a. Using Cis and Trans Prefixes

- **Cis-trans** prefixes are used to name alkenes with **two identical groups on both carbons of the double bond.**

- **General structures: cis-trans:**

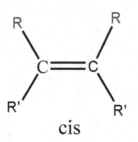

cis

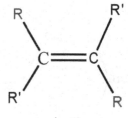

trans

Ex:

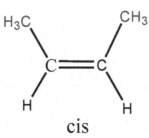

cis

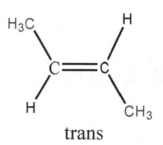

trans

b. Using E and Z Prefixes

- The **E and Z** prefixes are used to name diastereomers that have **three or four different groups bonded to both carbons of the double bond. Assignments are based on Cahn-Ingold-Prelog priority rules!**

- Review Unit 7: R and S configurations.

- E = Entgegen = opposite side
- Z = Zusammen = same side = **Z**ame **Z**ide.

- **General structures: E-Z**

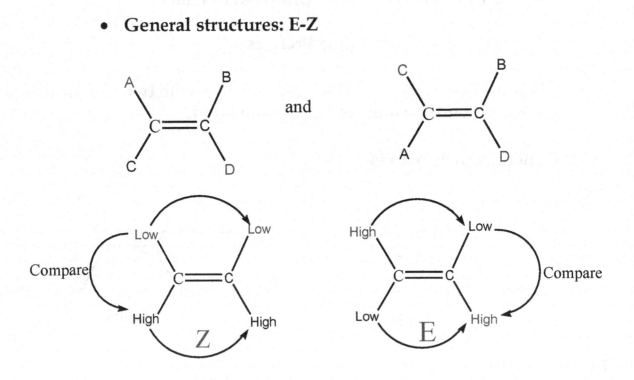

and

Ex: Using E/Z prefixes, name the following:

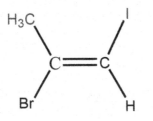

and

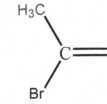

- **Read pages** _____ - _____.

- **Do Problems on page** _____.

5. ALKENES WITH OH GROUPS

- **Note: The OH group has a higher priority over the double bond. However, the parent chain takes the name of the corresponding alkene without the e.**

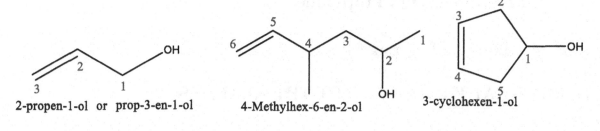

2-propen-1-ol or prop-3-en-1-ol   4-Methylhex-6-en-2-ol   3-cyclohexen-1-ol

6. COMMON NAMES FOR ALKENES

a. Alkenyl Groups:

$CH_2=$ methylene

$CH_2=CH-$ vinyl

$CH_2=CH-CH_2-$ allyl

Ex:

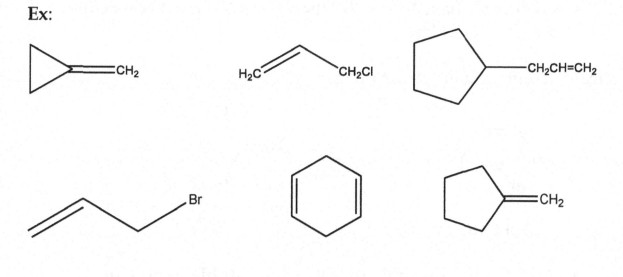

- **Read page _____ .**

b. Common Names

$$CH_2{=}CH_2 : \text{Ethylene}$$

$$CH_3CH{=}CH_2 : \text{Propylene}$$

- See Fig. _____, page _____.

D. PHYSICAL PROPERTIES OF THE ALKENES

1. INTRODUCTION

- In general, alkenes are **nonpolar** substances that have weak VWF between their molecules. As a result, their physical properties are similar to nonpolar alkanes.

2. MP + BP

- **Alkenes have low MP and BP. In general, BP increases with increasing molar mass because of increasing surface area.**

- **cis alkenes have higher BP than trans isomers because they are slightly polar.**

Ex: *cis*-2-butene and *trans*-2-butene

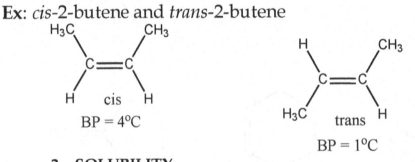

3. SOLUBILITY

- **Alkenes are insoluble in water, but soluble in organic solvents.**

- **Do problem on page _____.**

- **Which one of the following has the lowest BP? Explain**

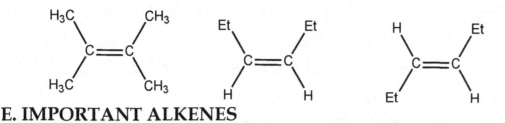

E. IMPORTANT ALKENES

- **Read pages _____ - _____.**

  ## 1. ETHYLENE AND POLYETHYLENE (PLASTICS)

- **Precursors of several products.**

  ## 2. LIPIDS

  ### a. Fatty Acids = Long chain carboxylic acids. Can be saturated or unsaturated.

- **See Fig. _____, page _____.**

- **See Table _____, page _____.**

  Lipids = Triacylglycerols (TAG)

- **Hydrolysis of triacylglycerols → Fatty acids**

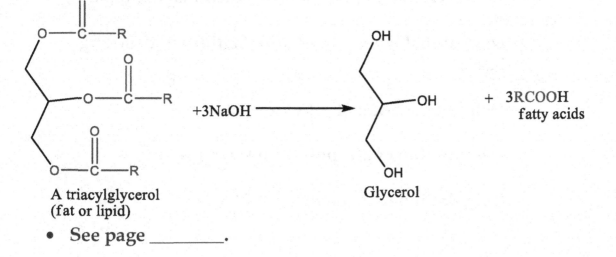

A triacylglycerol
(fat or lipid)

Glycerol

- **See page _____.**

Ex: Saturated and unsaturated fatty acids

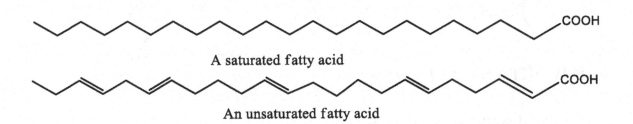

A saturated fatty acid

An unsaturated fatty acid

- As the number of double bonds in the R chain increases, MP of fatty acids decreases (due to smaller surface area).

    b. Fats + Oils:

- Fats = unsaturated TAG containing FA with few number of double bonds and high MP (larger surface area)

- Oils = unsaturated TAG with a relatively high number of double bonds and low MP (smaller surface areas).

F. SYNTHESIS OF ALKENES

1. INTRODUCTION

- There are several ways (**elimination reactions**) of preparing alkenes:
    o **Dehydrohalogenation of alkyl halides with strong bases.**

    o **Dehydrogenation of alkyl tosylates.**

    o **Dehydration of alcohols with strong acids.**

    o **Dehydration of alcohols with POCl$_3$.**

2. E2 DEHYDROHALOGENATION OF ALKYL HALIDES WITH STRONG BASES (OH⁻, RO⁻, NH₂⁻, DBU, DBN)

- The E2 reaction proceeds as follows:

:B⁻

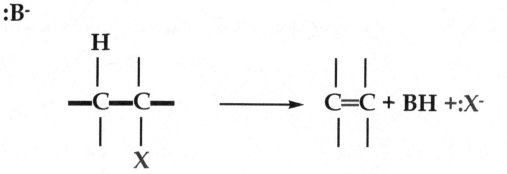

An alkene

Where **:B⁻ = a strong base = OH⁻, RO⁻, NH₂⁻, DBN, DBU**

- Note: The reaction is stereoselective and regioselective and follows Zaitsev's rule.

Ex:

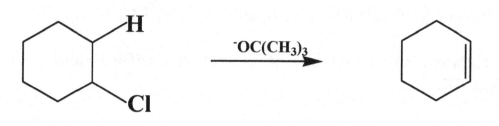

3. E2 DEHYDROGENATION OF ALKYL TOSYLATES

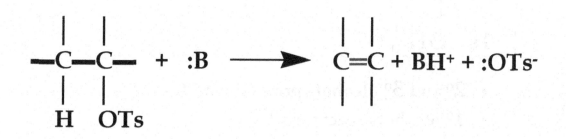

Ex:

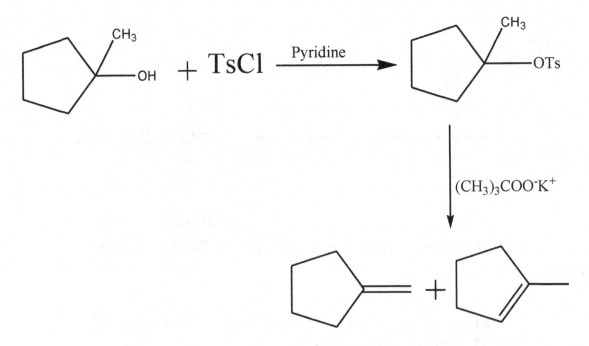

4. DEHYDRATION OF ALCOHOLS WITH H_2SO_4, TsOH, or H_3PO_4 IN THF

- This is a β elimination reaction. See Unit 12.

- Catalysts: strong acids: H_2SO_4, p-toluenesulfonic acid (TsOH).

General reaction:

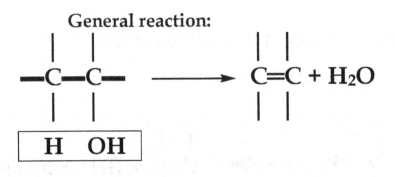

- 2° and 3° alcohols proceed via E1.
- 1° alcohols react thru **E2.**

- The reaction follows Zaitsev's rule. The reaction is stereoselective and regiospecific.

Ex:

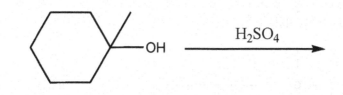

5. E2 DEHYDRATION OF ALCOHOLS USING POCl$_3$ IN PYRIDINE

POCl$_3$ = phosphorus oxychloride

- **General Reaction: See Unit 12.**

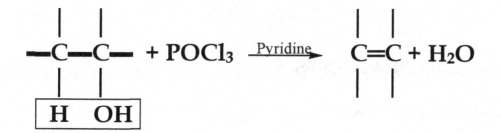

Ex:

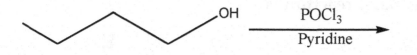

6. SYNTHESIS OF THE ALKENE: A SUMMARY

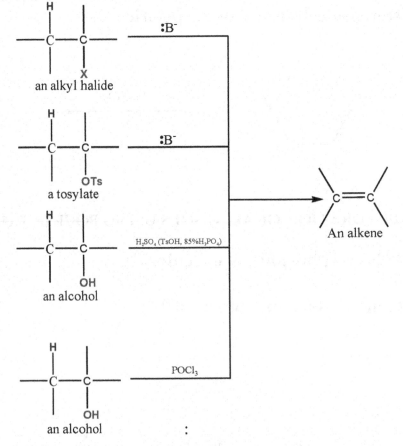

G. REACTIONS OF ALKENES

1. INTRODUCTION

- Unlike the alkanes, **alkenes** are **very reactive** because of their **unsaturations.** They are **electron-rich compounds** that react as **nucleophiles.** In general, they undergo **addition reactions.**

- **The general reaction is:**

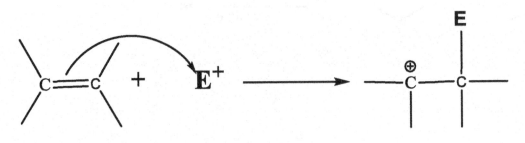

- See Fig. _____, page _____, a Summary of Addition reactions.

2. ADDITION OF HALOGENS TO THE ALKENES: HALOGENATION

a. Introduction

- Br_2 and Cl_2 add readily.
- F_2 is too reactive and difficult to control
- I_2 is difficult to add. It does react with few alkenes.
- The reaction is antistereochemistry: anti addition: the 2 halogen atoms add on opposite sides.

b. General Reaction

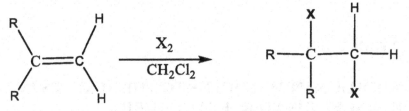

- Mechanism: The reaction proceeds through a bridged halonium ion:

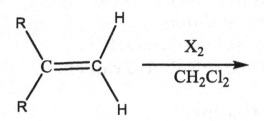

Ex:

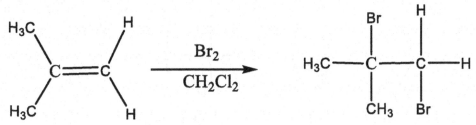

371

c. Stereochemistry of Halogenation

- The reaction is **stereospecific**. We have a **trans addition.**

Ex:

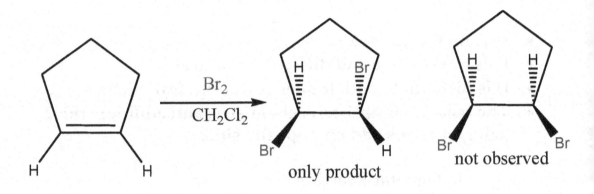

only product

not observed

- Do problems on pages _____ - _____.

3. ADDITION OF HALOGENS TO THE ALKENES IN THE PRESENCE OF WATER: HALOHYDRINS

a. Introduction

- Only Cl_2 and Br_2 react.
- Halohydrins products=1,2 alcohols.
- The reaction = anti addition.
- X adds to the least substituted carbon.
- OH adds to the most substituted carbon.

b. General Reaction

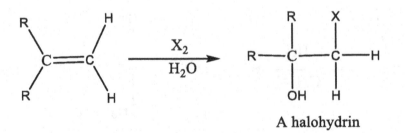

A halohydrin

c. Mechanism: Reaction Proceeds in 2 Steps

- The **-OH group** goes to the **most substituted** carbon of the double bond.

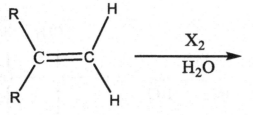

Ex:

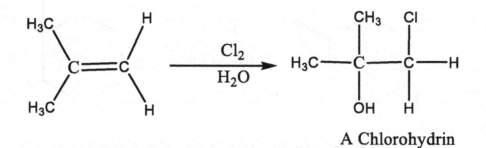

A Chlorohydrin

d. Bromohydrins from NBS: Indirect Bromination

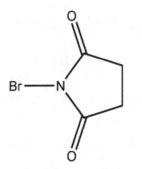

N-bromosuccinimide

- **General Reaction:**

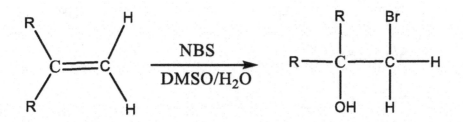

- **Note: NBS is the Br source. The -OH goes to the most substituted C of the double bond.**

Ex:

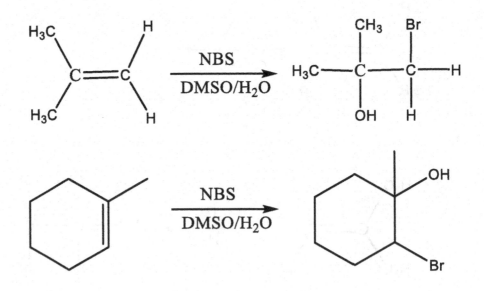

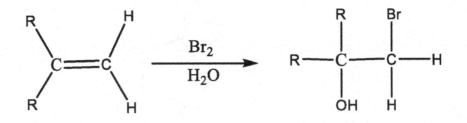

e. Bromohydrins from the Direct Reaction of Br₂.

- **The general reaction is:**

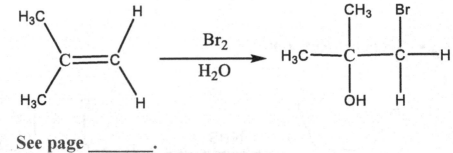

Ex:

- See page _____.

- See Table _____, page _____.

4. ADDITION OF OTHER NUCLEOPHILES TO THE ALKENES IN THE PRESENCE OF X₂

a. Introduction

- One can add ROH, MX (M= K, Na, etc. X = Cl, Br) in the presence of X_2 (X=Cl, Br, I).

b. The general reaction is:

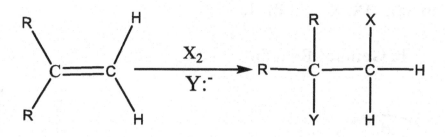

c. Some Examples

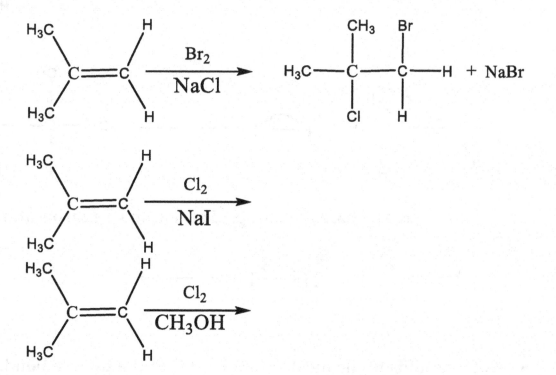

- Note: Y:⁻ adds to the most substituted C of the double bond.

375

5. ELECTROPHILIC ADDITION OF HX TO THE ALKENES: HYDROHALOGENATION

a. Introduction

- Syn and anti additions occur.

- Overall reaction is exothermic.

- Can add HX: X= Cl, Br, I.

b. General Reaction

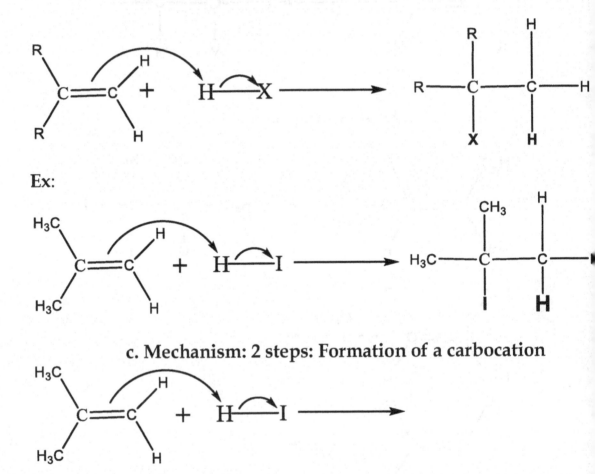

Ex:

c. Mechanism: 2 steps: Formation of a carbocation

- Note: X:⁻ adds to the most substituted C of the double bond.
- See Fig. _____, page ____. Do Problems, pages _____ - _____.

d. Markovnikov's Rule

- It applies to unsymmetrically substituted alkenes.
- In adding HX to unsymmetrical alkenes, the H always goes to the carbon of the double bond with the most number of hydrogens. In other words, X:⁻ adds to the most substituted C of the double bond.

- Read pages _____ - _____.

Ex:

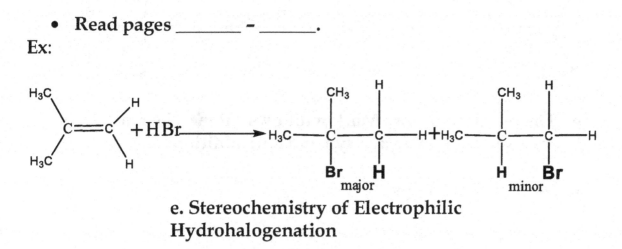

e. Stereochemistry of Electrophilic Hydrohalogenation

- Since the X⁻ can attack the carbocation intermediate on two sides, both syn and anti additions are observed and racemization occurs.

Ex:

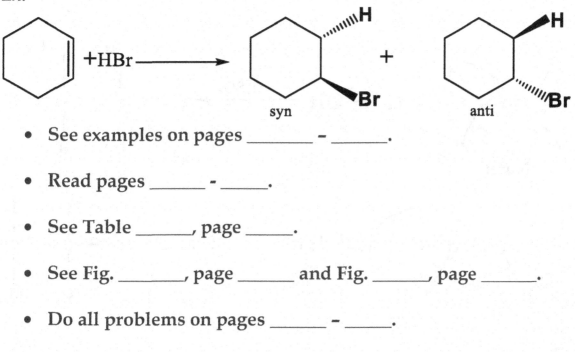

- See examples on pages _____ - _____.

- Read pages _____ - _____.

- See Table _____, page _____.

- See Fig. _____, page _____ and Fig. _____, page _____.

- Do all problems on pages _____ - _____.

6. ELECTROPHILIC ADDITION OF H_2O TO THE ALKENES: HYDRATION

a. Direct Method Using H_2SO_4 or H_3PO_4

- **General Reaction:**

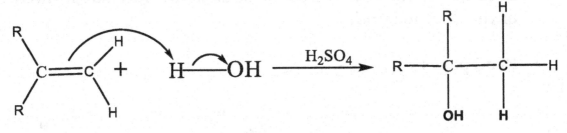

- **The reaction follows Markovnikov's rule➡ The reaction is regioselective. We have syn and anti addition.**

- **Mechanism: 3 steps**

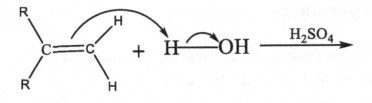

Ex:

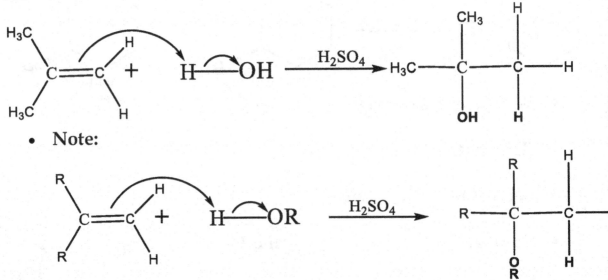

- **Note:**

Ex:

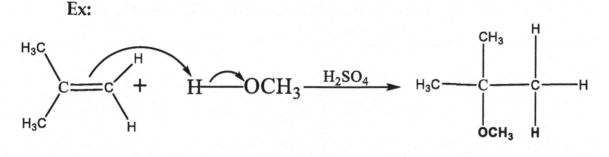

- Read pages _____ - _____.

- Do problems on page _____.

      b. Indirect Method Using BH₃: Hydroboration: An Oxidation Reaction

- The general reaction:

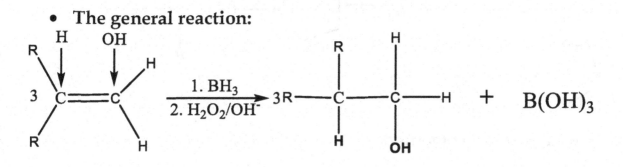

- **See mechanism on page _____.**
- The reaction is a syn addition: H and OH add to the same side.
- The reaction is nonMarkovnikov➔ The H adds to the carbon with the least number of hydrogens.
- The stereochemistry of the reactant alkene is conserved: Cis ➔ Cis; Trans➔ Trans.

Ex:

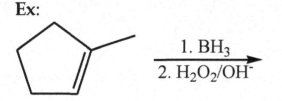

- Do problems on page _____.

- Read pages _____ - _____.

c. Addition of Water to the Alkenes: Direct vs. Indirect Hydration

- See Table _____, page _____.

- Do problems on pages _____ – _____.

Ex:
- **The direct method**

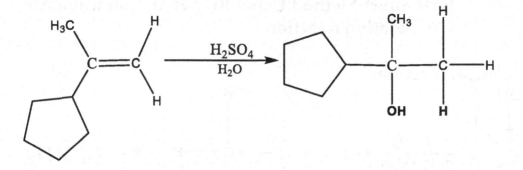

- **The indirect method**

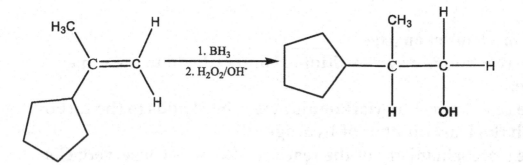

7. REACTIONS OF THE ALKENES: A SUMMARY

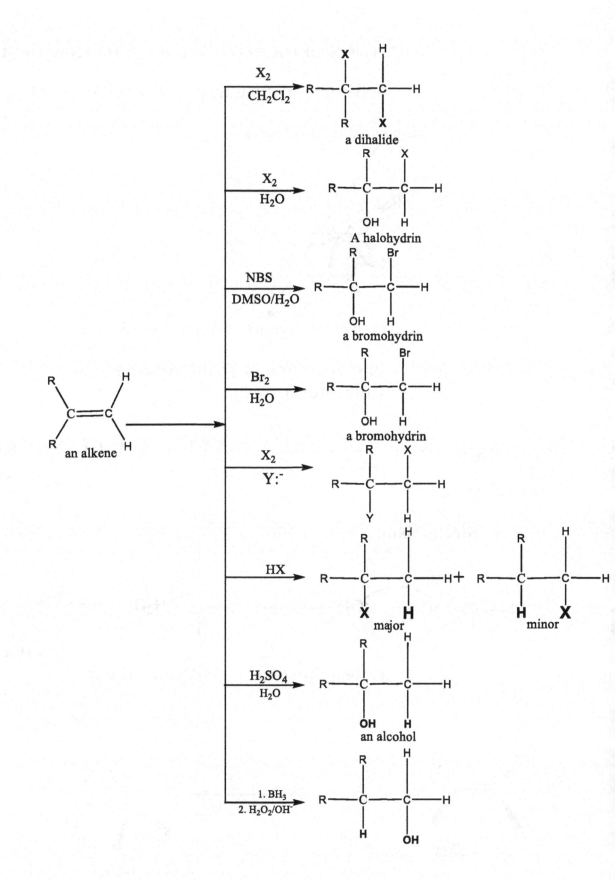

8. ADDITION OF DICHLOROCARBENE TO AN ALKENE

a. Structure of a Carbene: A Review (See Unit 8)

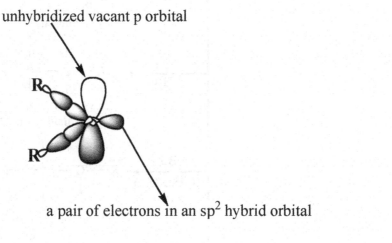

unhybridized vacant p orbital

R

R

a pair of electrons in an sp^2 hybrid orbital

b. Preparation of Dichlorocarbene from Chloroform

$$CHCl_3 \xrightarrow[\text{or KOH}]{KOC(CH_3)_3} :CCl_2 + (CH)_3COH + KCl$$

chloroform

- **Mechanism:**

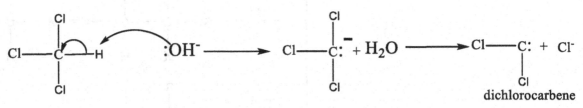

c. Reaction of :CCl₂ with an Alkene

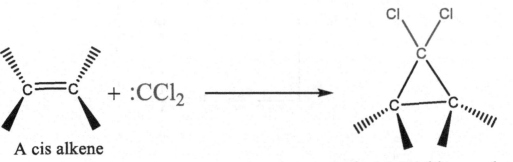

A cis alkene

+ :CCl₂ ⟶

A cis-1,1-Dichlorocyclopropane

Ex:

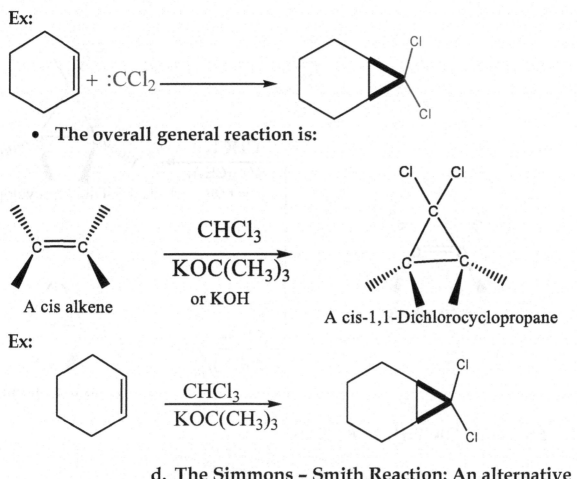

• The overall general reaction is:

A cis alkene
$$\xrightarrow[\text{or KOH}]{\begin{array}{c}\text{CHCl}_3\\ \hline \text{KOC(CH}_3)_3\end{array}}$$
A cis-1,1-Dichlorocyclopropane

Ex:

$$\xrightarrow[\text{KOC(CH}_3)_3]{\text{CHCl}_3}$$

d. The Simmons – Smith Reaction: An alternative to Dichloro Cyclic Products

• The catalyst is CH_2I_2/Zn(Cu) in ether:

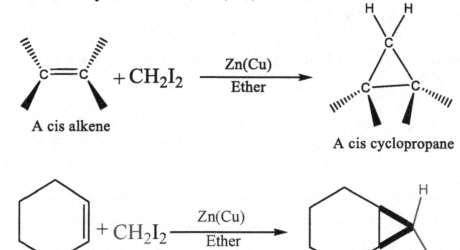

A cis alkene $+ CH_2I_2 \xrightarrow[\text{Ether}]{\text{Zn(Cu)}}$ A cis cyclopropane

Ex:

$+ CH_2I_2 \xrightarrow[\text{Ether}]{\text{Zn(Cu)}}$

9. PREPARATION OF CYCLOPROPANES FROM ALKENES: A SUMMARY

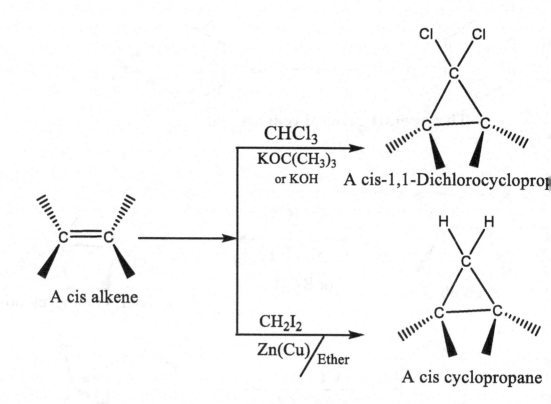

$$CHCl_3$$
$$KOC(CH_3)_3$$
or KOH

A cis-1,1-Dichlorocyclopro[

A cis alkene

$$CH_2I_2$$
$$Zn(Cu)/Ether$$

A cis cyclopropane

H. TESTING FOR THE ALKENES

1. INTRODUCTION

- There are several **qualitative** tests that can be performed to prove whether or not an unknown sample is an alkene. Here, we focus on **3 tests**.

2. THE BROMINE WATER (DILUTE) TEST: AN ADDITION REACTION

- An alkene is **colorless.**
- **Bromine is red-brown.**

- **The general reaction is:**

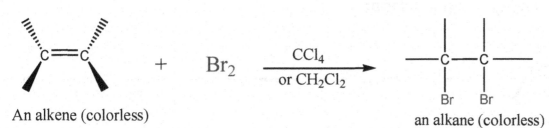

An alkene (colorless) $+$ Br_2 $\xrightarrow[\text{or } CH_2Cl_2]{CCl_4}$ an alkane (colorless)

- **Note: A positive test results in the decolorization of the bromine red-brown color. An alkane is used as a control.**

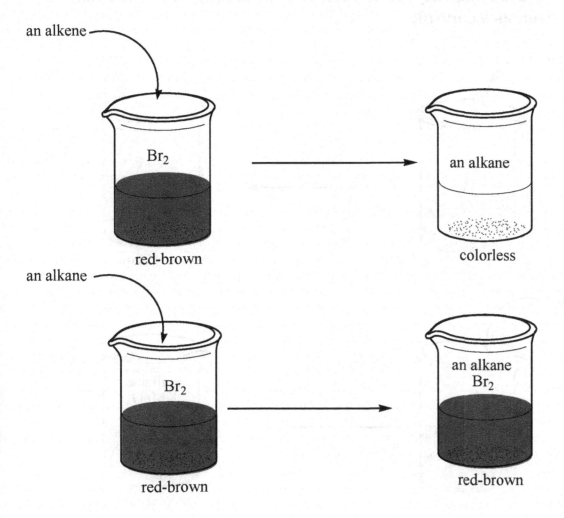

3. THE BAEYER TEST: AN OXIDATION REACTION

- **An alkene is colorless.**
- **$KMno_4$ is purple; the product MnO_2 is brown.**

- **The general reaction:**

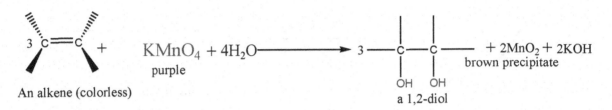

An alkene (colorless)

$KMnO_4$ + $4H_2O$ 3 — C — C — + $2MnO_2$ + $2KOH$

purple

OH OH

a 1,2-diol

brown precipitate

- **Note: A positive test results in the color change from clear purple ($KMnO_4$) to a brown precipitate (MnO_2). An alkane is used as a control.**

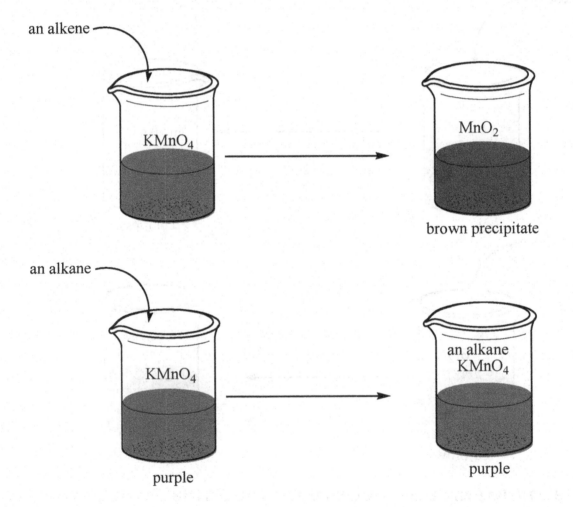

an alkene

$KMnO_4$

MnO_2

brown precipitate

an alkane

$KMnO_4$

purple

an alkane
$KMnO_4$

purple

4. THE SULFURIC ACID TEST: AN ADDITION REACTION

- The reagent is concentrated H_2SO_4.

- The general reaction is:

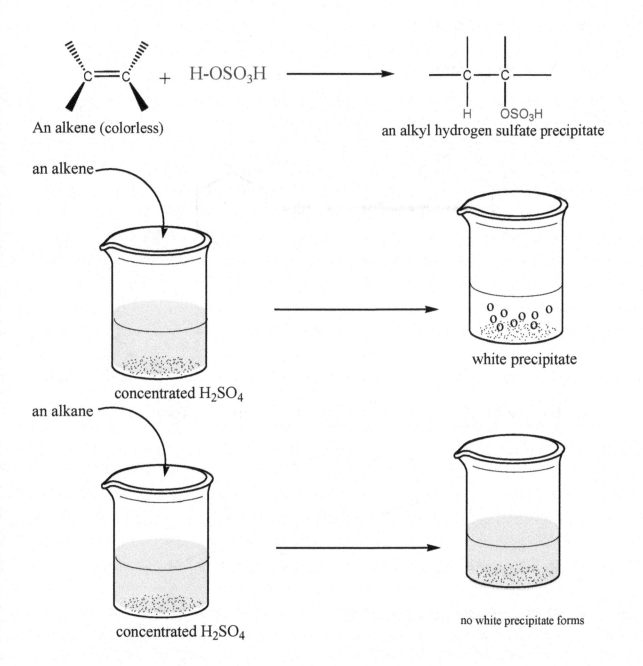

An alkene (colorless)

$H\text{-}OSO_3H$

an alkyl hydrogen sulfate precipitate

an alkene

concentrated H_2SO_4

white precipitate

an alkane

concentrated H_2SO_4

no white precipitate forms

- **Note: A positive test results in the formation of white crystals. An alkane is used as a control.**

I. INTRODUCTION TO ORGANIC SYNTHESIS

- 3 things are needed:

  A, B, and a catalyst.

Ex:

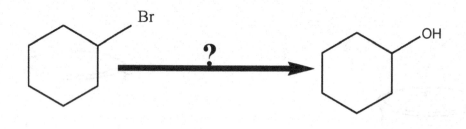

- How would you synthesize the following compound?

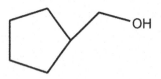

- **Read pages _____ - _____.**

- **Do Problem _____, page _____.**

J. A SUMMARY OF SOME SELECTED REACTIONS

1. SYNTHESIS OF THE ALKENE: A SUMMARY

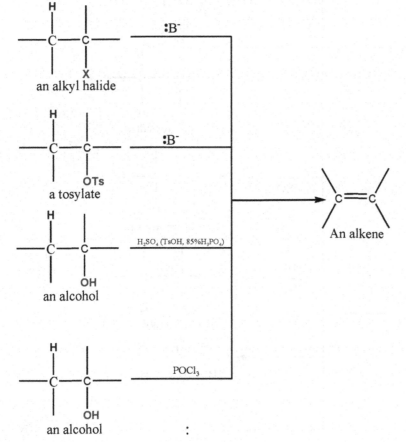

an alkyl halide

a tosylate

an alcohol

an alcohol

2. REACTIONS OF THE ALKENES: A SUMMARY

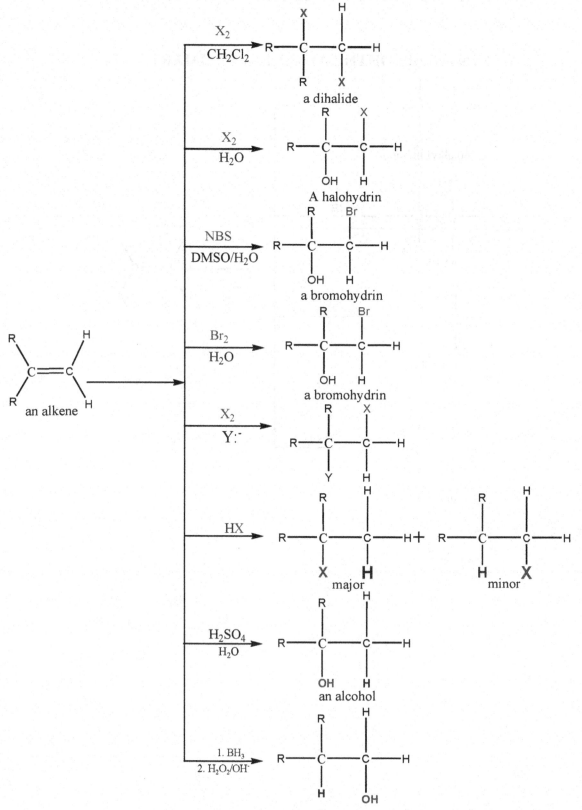

3. PREPARATION OF CYCLOPROPANES FROM ALKENES: A SUMMARY

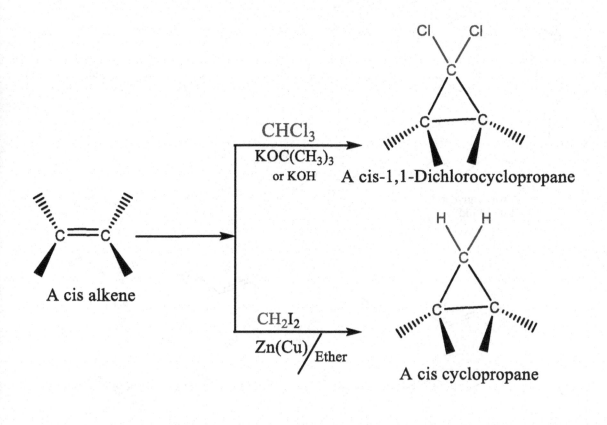

A cis-1,1-Dichlorocyclopropane

A cis cyclopropane

A cis alkene

- See Key Concepts on pages _____ – _____.

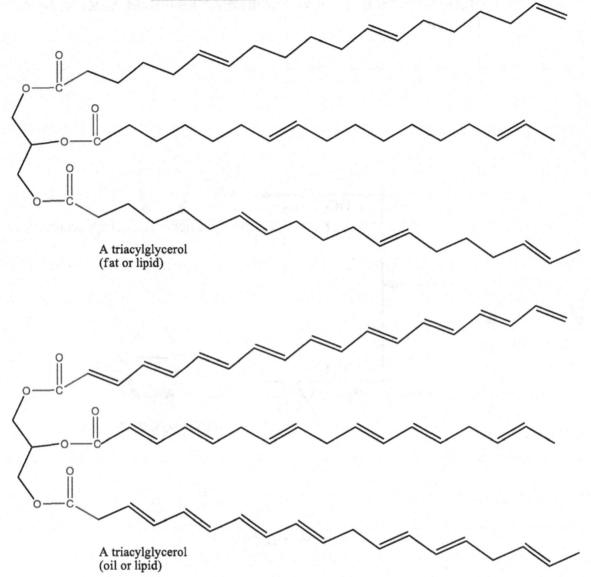

A triacylglycerol
(fat or lipid)

A triacylglycerol
(oil or lipid)

OCHEM I UNIT 15: ALKYNES

A. INTRODUCTION

1. DEFINITION

- **Alkynes** are **unsaturated hydrocarbons** that contain at least one **C-C triple bond.**

2. GENERAL MOLECULAR FORMULA OF ALKYNES

$$C_nH_{2n-2}$$

n = **Total number of carbon atoms in molecule.**
- **Note: Have a homologous series.**

Ex: n = 1

n = 2

n = 6

n = 10

3. TYPES OF ALKYNES: 4 types

a. Terminal Alkynes

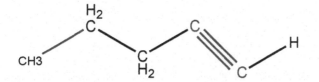

b. Internal Alkynes

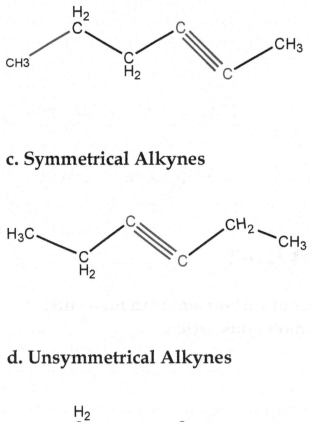

c. Symmetrical Alkynes

d. Unsymmetrical Alkynes

4. PROPERTIES OF THE C-C TRIPLE BOND

- The **-C≡C-** is a very rigid bond. It **cannot be rotated** like a single bond.

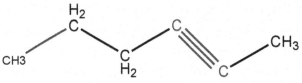

B. NAMING ALKYNES

1. IUPAC NAMES

- Same rules as alkenes, except the names of alkynes end in –yne (or -diyne).

n	molecular formula	name	structure
2	C_2H_2	Eth-1-yne	$H-C\equiv C-H$
3	C_3H_4	Prop-1-yne	$CH_3C\equiv C-H$
4	C_4H_6	But-1-yne	$CH_3CH_2C\equiv C-H$
4	C_4H_4	But-1-en-3-yne	$CH_2=CHC\equiv C-H$

2. ISOMERISM IN ALKYNES

$n = 4$ $CH_3CH_2C\equiv C-H$ but-1-yne or 1-butyne
$CH_3C\equiv CCH_3$ but-2-yne or 2-butyne

3. POLYYNES

- Polyynes are alkynes with 2 or more triple bonds.

$H-C\equiv CCH_2CH_2CH_2C\equiv C-H$ hepta-1,6-diyne or 1,6-heptadiyne

4. COMMON NAMES
- Alkynes are named as **acetylenes**.

structure	IUPAC	common name
$H-C\equiv C-H$	ethyne	**acetylene**
$CH_3C\equiv C-H$	propyne	**methyl** acetylene
$CH_3C\equiv CCH_3$	2-butyne	**dimethyl** acetylene
$CH_2=CHC\equiv C-H$	1-buten-3-yne	**vinyl** acetylene

- **Note: The three bonds in a triple bond in structural formulas of alkynes are always written out (shown).**

5. ALKYNYL GROUPS

CH₃CH₂C≡C- 1-butynyl

HC≡C- ethynyl

Ex:

Ethynylcyclopropane

- Do problem on page_____.

- **Note: Small ring cycloalkynes are unstable. Cyclooctyne is the lowest cycloalkyne ever isolated. However, it is unstable.**

- Read page _____.

C. PHYSICAL PROPERTIES OF ALKYNES

- Like the alkanes and alkenes, alkynes are nonpolar. Therefore, they have only weak VWF between their molecules. As a result, they have physical properties similar to those of the other hydrocarbons (low BP, MP, insoluble in water, etc.)

D. IMPORTANT ALKYNES

- See Fig. _____, page _____.

- See page _____.

- **Acetylene**
- **Ethynylestradiol (looks like estradiol; prevents ovulation)**
- **norethindrone (looks like progesterone; induces the production of a thick mucus that prevents the sperm to reach the uterus) = birth control pill.**

- **RU 486 (Mifepristone):** "the morning after pill"; prevents implantation of a fertilized egg.
- **Plan B = levonoorgestrel.**
- **Histrionicotoxin: a poison from Amazonian frog.**

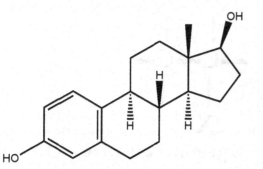

estradiol, a female hormone

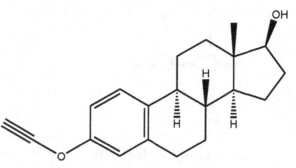

ethynylestradiol, a synthetic hormone

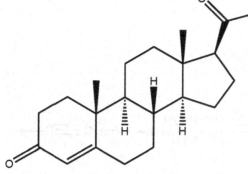

progesterone, a female hormone

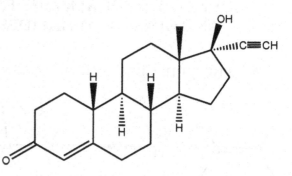

norethindrone, a synthetic hormone

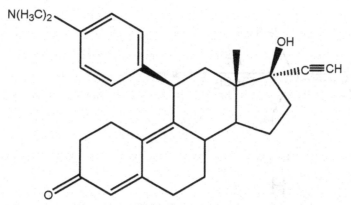

RU 486 (mifepristone), a synthetic hormone

- **Read pages _____ – _____.**

E. SYNTHESIS OF ALKYNES

1. VICINAL OR GEMINAL DIHALIDES

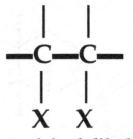

A vicinal dihalide

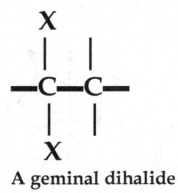

A geminal dihalide

2. SYNTHESIS OF ALKYNES FROM ALKENES VIA VICINAL AND GEMINAL DIHALIDES

- This general reaction proceeds in two steps as follows:

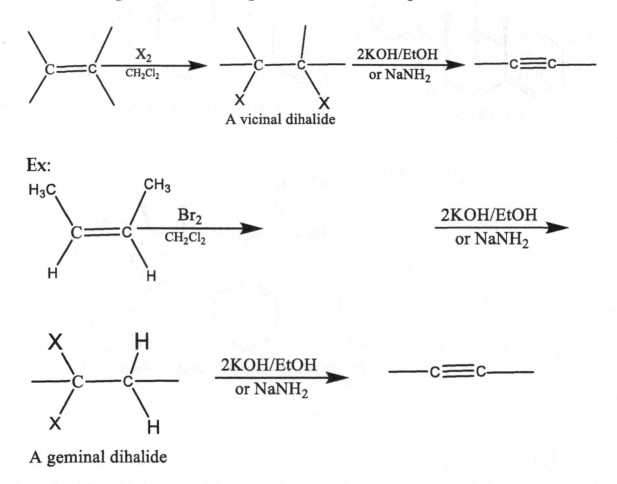

Ex:

Ex:

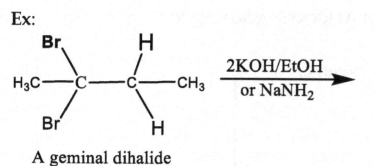

A geminal dihalide

- See page _____; See Unit _____, page _____.

3. SYNTHESIS OF ALKYNES FROM VINYLIC HALIDES

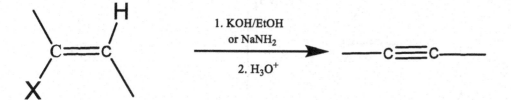

Ex:

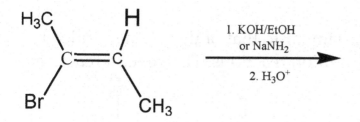

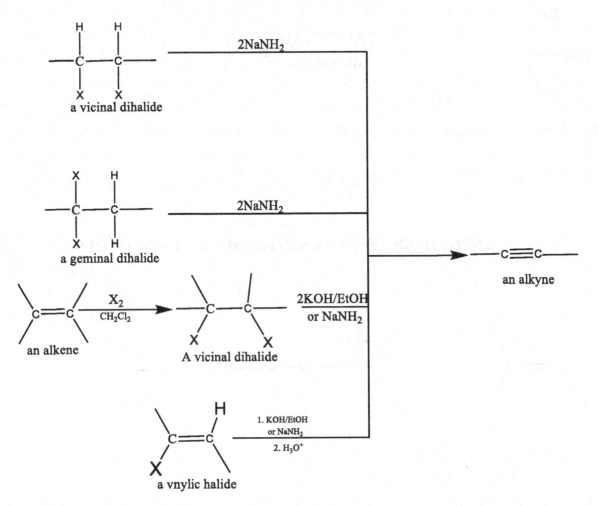

F. REACTIONS OF ALKYNES

1. INTRODUCTION

The chemistry of the alkynes is similar to that of the alkenes. Indeed they act like **nucleophiles** in addition reactions. The general reaction:

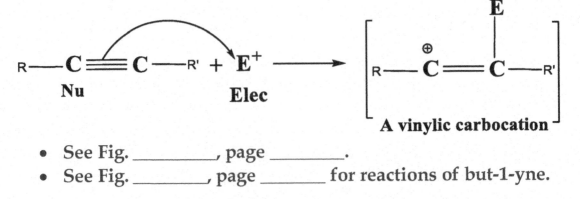

- See Fig. _____, page _____.
- See Fig. _____, page _____ for reactions of but-1-yne.

400

2. ADDITON OF HX (X = Cl, Br, I) TO THE ALKYNES: HYDROHALOGENATION

- **The reaction follows Markovnikov's rule**
- **A trans-product is first produced followed by a geminal dihalide.**

- **The general reaction is:**

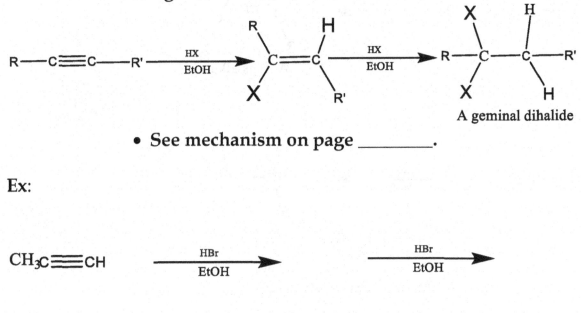

A geminal dihalide

- **See mechanism on page _____.**

Ex:

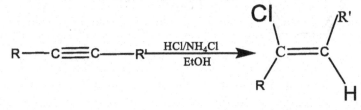

$CH_3C\equiv CH$ $\xrightarrow[\text{EtOH}]{\text{HBr}}$ $\xrightarrow[\text{EtOH}]{\text{HBr}}$

- **Note: HCl can also be added to the alkynes as follows:**

R—C≡C—R' $\xrightarrow[\text{EtOH}]{\text{HCl/NH}_4\text{Cl}}$

Ex:

$CH_3C\equiv CCH_3$ $\xrightarrow[\text{EtOH}]{\text{HCl/NH}_4\text{Cl}}$

3. ADDITON OF X_2 (X = Cl, Br) TO THE ALKYNES: HALOGENATION

- **The reaction is a trans-addition. The general reaction is:**

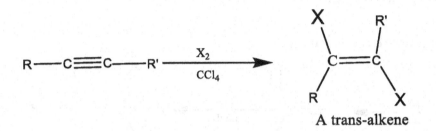

A trans-alkene

or

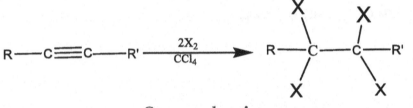

- See mechanism on page _____.

- Do problems on page _____ - _____.

Ex:

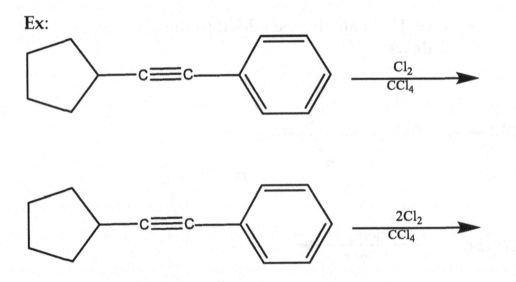

4. ADDITION OF WATER TO THE ALKYNES: HYDRATION

   a. **Introduction: 2 ways of adding water to the triple bond**

   - Direct hydration: Uses a mercury(II) sulfate catalyst.
   - A Markonikov product is obtained.
   - Indirect hydration: Hydroboration.
   - A nonMarkovnikov product is obtained after tautomerization.

   b. **Mercuration of Alkynes**

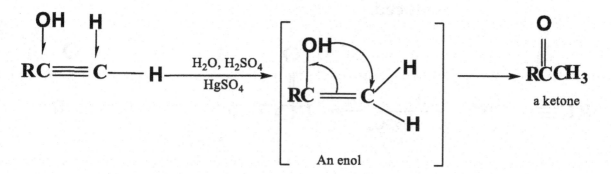

   - **Enol and ketone = tautomers**

   - **See mechanism on page _____.**

Ex:

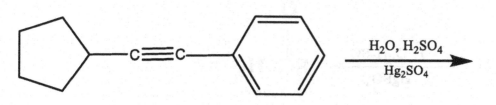

   - **Do example and problems on page _____.**

c. Products of Mercuration

   i. Internal symmetrical alkynes: one ketone is produced.

$$RC\equiv CR \xrightarrow[\text{HgSO}_4]{\text{H}_2\text{O, H}_2\text{SO}_4} RC\overset{\overset{\displaystyle O}{\|}}{C}CH_2R$$

a ketone

Ex:

$$CH_3C\equiv CCH_3 \xrightarrow[\text{HgSO}_4]{\text{H}_2\text{O, H}_2\text{SO}_4}$$

   ii. Unsymmetrical internal alkynes: two ketones are produced.

$$RC\equiv CR' \xrightarrow[\text{HgSO}_4]{\text{H}_2\text{O, H}_2\text{SO}_4} RC\overset{\overset{\displaystyle O}{\|}}{C}CH_2R' \ + \ RCH_2\overset{\overset{\displaystyle O}{\|}}{C}R'$$

2 ketones

Ex:

$$CH_3C\equiv CCH_2CH_3 \xrightarrow[\text{HgSO}_4]{\text{H}_2\text{O, H}_2\text{SO}_4}$$

   iii. Terminal alkynes: one ketone is produced.

$$RC\equiv CH \xrightarrow[\text{HgSO}_4]{\text{H}_2\text{O, H}_2\text{SO}_4} RC\overset{\overset{\displaystyle O}{\|}}{C}CH_3$$

A ketone

Ex:

$$CH_3C\equiv CH \xrightarrow[\text{HgSO}_4]{\text{H}_2\text{O, H}_2\text{SO}_4}$$

d. Hydroboration: Indirect Hydration

- This is a nonMarkovnikov oxidation reaction. The general reaction is:

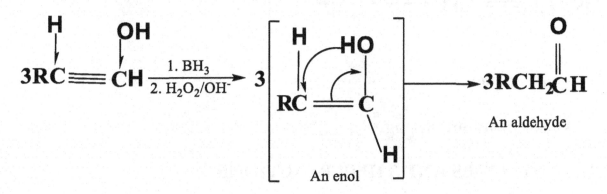

Ex:

$$3CH_3C \equiv CH \xrightarrow[\text{2. } H_2O_2/OH^-]{\text{1. } BH_3/THF}$$

e. Products in Alkyne Hydroboration

i.   **Internal symmetrical alkynes** ⟶ one ketone.

Ex:

$$3CH_3C \equiv CCH_3 \xrightarrow[\text{2. } H_2O_2/OH^-]{\text{1. } BH_3/THF}$$

ii.   **Internal unsymmetrical alkynes** ⟶ 2 ketones.

Ex:

$$3CH_3C \equiv CCH_2CH_3 \xrightarrow[\text{2. } H_2O_2/OH^-]{\text{1. } BH_3/THF}$$

iii. Terminal Alkynes ⟶ one aldehyde

Ex:

$$3CH_3C \equiv CH \xrightarrow[\text{2. } H_2O_2/OH^-]{\text{1. } BH_3/TH_F}$$

- Read pages _____ - _____.

- Do problem on page _____.

G. ACETYLIDES AND THEIR REACTIONS

1. INTRODUCTION

- Terminal alkynes are **weakly acidic (pKa = 25)**. Indeed, they are more acidic than alkanes (pKa = 60) and alkenes (pKa = 44). Therefore, they can react with bases.

2. FORMATION OF ACETYLIDES

$$RC \equiv CH \xrightarrow[\text{liquid } NH_3]{NaNH2} RC \equiv C\colon^- Na^+ + NH_3$$

An acetylide anion

Ex:

$$CH_3C \equiv CH \xrightarrow[\text{liquid } NH_3]{Na^+NH_2^-} CH_3C \equiv C\colon^- Na^+ + NH_3$$

An acetylide anion

3. REACTIONS OF THE ALKYNES: A SUMMARY

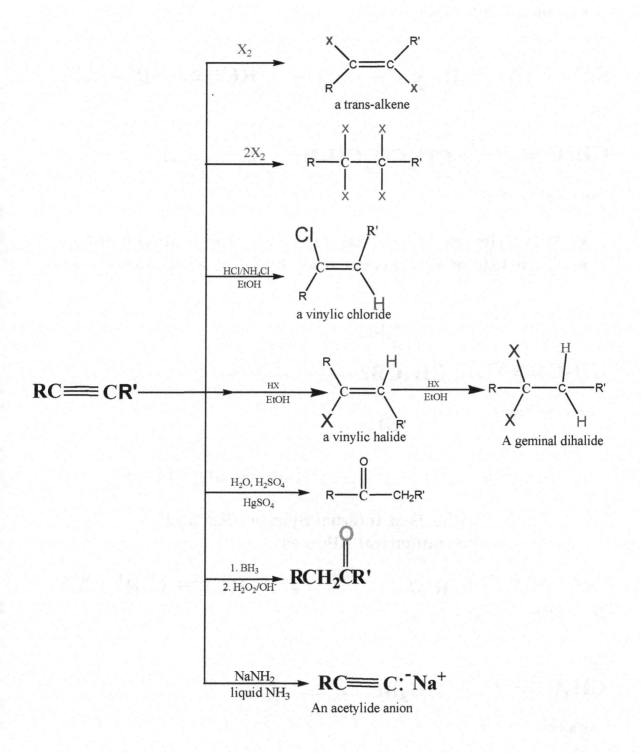

4. ACETYLIDES AS NUCLEOPHILES

a. Introduction

- The general reaction:

$$RC\equiv C:^- + R'-X \xrightarrow{\text{SN2}} RC\equiv C-R' + X:^-$$

Ex:

$$CH_3C\equiv C:^- + CH_3CH_2CH_2Br \longrightarrow$$

An acetylide anion

- Note: -The reaction is fastest for CH_3X and 1° alkyl halides.
- E2 (instead of SN2) occurs when bulky alkyl halides are used.

Ex:

$$CH_3C\equiv C:^- + CH_3\overset{\displaystyle CH_3}{\underset{\displaystyle CH_3}{\overset{|}{\underset{|}{C}}}}Br \longrightarrow$$

An acetylide anion

- See pages _____ - _____ and do all problems.

b. Synthesis of Internal Symmetrical and Unsymmetrical Alkynes

$$RC\equiv C:^- + R'-X \xrightarrow{\text{SN2}} RC\equiv C-R' + X:^-$$

R = R': symmetrical alkyne

Ex:

$$CH_3C\equiv C:^- + CH_3Br \longrightarrow$$

An acetylide anion

$$RC\equiv C\overset{-}{:} + R'\text{-}X \xrightarrow{SN2} RC\equiv C\text{-}R' + X\overset{-}{:}$$

R different from **R':** unsymmetrical alkyne

Ex:

$$CH_3C\equiv C\overset{-}{:} + CH_3CH_2CH_2Br \longrightarrow$$

An acetylide anion

5. OPENING OF EPOXIDES BY ACETYLIDE ANIONS

The general reaction is:

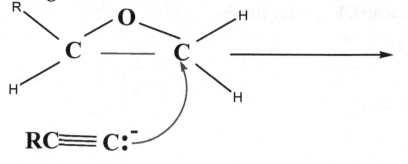

Ex:

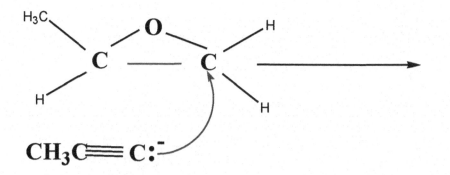

- See page _____.

- **Note: The attack occurs at the least substituted carbon.**

H. CHEMICAL SYNTHESIS REVISITED

- **Question: How do you prepare B from A?**

Find all possible catalysts/reagents.

Ex:

$$CH_3CH_2CH_2C \equiv CH \longrightarrow CH_3(CH_2)_6CH_3$$

$\uparrow$ 5 carbons $\uparrow$ 8 carbons

- **Note: Work backward: Retrosynthesis.**

- **Read pages** _____ – _____.

- **Do problems on page** _____.

I. SOME SELECTED CHEMICAL REACTIONS

1. SYNTHESIS OF ALKYNES: A SUMMARY

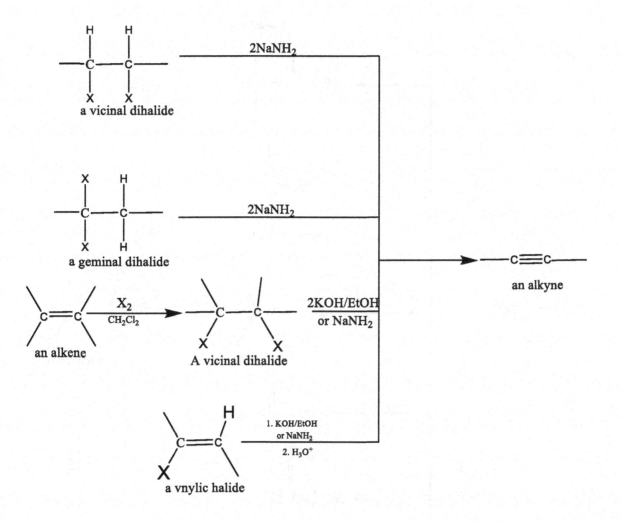

2. REACTIONS OF THE ALKYNES: A SUMMARY

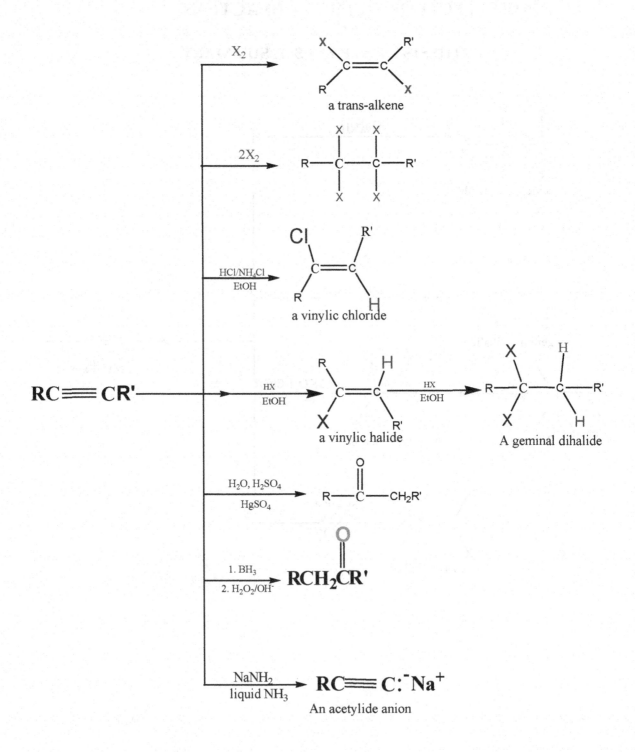

- **See Key Concepts and Summary of Reactions on page** _____ - _____.

OCHEM I UNIT 16: REDOX REACTIONS IN OCHEM

A. INTRODUCTION: AN OVERVIEW ON REDOX REACTIONS

1. SOME BASIC DEFINITIONS

- A reducing agent (*reductant*) is a reactant that loses (or donates) electrons to a second reactant called an oxidizing agent (*oxidant*) in an electron transfer reaction.

- Note: The reducing agent becomes oxidized at the completion of the reaction.

- Note: A process in which electrons are lost is called an oxidation reaction.

- On the other hand, an oxidizing agent (oxidant) is a reactant that gains (or accepts) electrons from another reactant called a reducing agent (reductant) in an electron transfer reaction.

- Note: The oxidizing agent becomes reduced at the completion of the reaction.

- Note: A process in which electrons are gained is called a reduction reaction.

- Note: Oxidation and reduction occur in tandem (meaning happen together) anytime there is an electron transfer during a reaction. Putting half of each word together, the word redox is derived.

- So, a redox reaction is a reaction in which one or more electrons are transferred from the reducing agent (most generous one in electrons) to the oxidizing agent (stingiest in electrons).

- In short, Reducing agent + Oxidizing agent = Redox Reaction

- The following example is a good illustration of a redox reaction:

  o $Zn + Cu^{2+} \longrightarrow Zn^{2+} + Cu$

- In this reaction:

- Zn = reducing agent = losing 2 electrons to Cu^{2+}

- Cu^{2+} = oxidizing agent = gaining 2 electrons from Zn

2. REDOX REACTIONS IN OCHEM

- Now, let's apply the concepts described above to organic reactions. Unfortunately, redox reactions in OCHEM are not that "*straightforward*". Redox reactions here occurs through a hydride (:H⁻) and/or a proton (H⁺). In other words, redox reactions occur when two electrons are transferred through a C-H bond. A loss of :H⁻ (2e⁻) from a carbon corresponds to a loss of electron density on that carbon. This process is an oxidation (loss of electrons). Conversely, a gain of :H⁻ (2e⁻) by a carbon is a gain of electron density for that carbon. This process is a reduction (gain of electrons). The dehydrogenation of an alcohol reaction to give a ketone can be used to illustrate Ochem redox reactions:

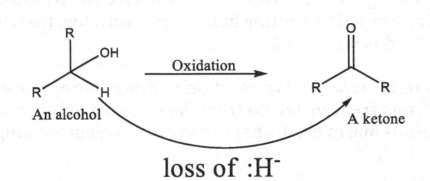

loss of :H⁻

- **The hydrogenation of an alkene to an alkane (addition of H₂) is an example of organic reduction reaction:**

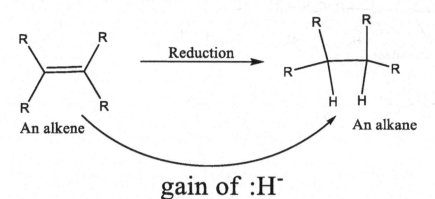

$$\text{gain of :H}^-$$

- **Indeed, in this reaction, the electron density on one of the carbons of the double bond increases due to the incoming :H⁻ (hydride) from H₂.**

3. **REDOX REACTIONS IN OCHEM: A SUMMARY**

- **Reduction = Addition of a hydride :H⁻(and/or H⁺)**

- **Oxidation = Removal of a hydride :H⁻(and/or H⁺)**

- **Note: Anytime you take away a hydride, oxidation will occur because you have a decrease in electron density.**

- **Note: The more the CO bonds, the more oxidized the compounds. The following compounds are ranked in increasing order of degree of oxidation:**

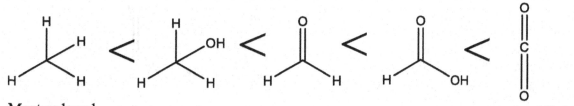

Most reduced   Most oxidized

or

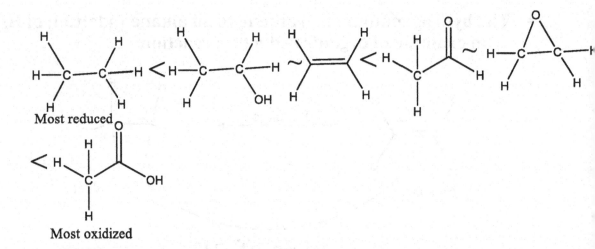

Most reduced

Most oxidized

- **Question: Is the following substitution reaction a redox reaction?**

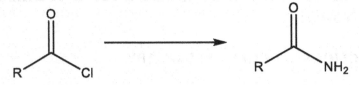

B. REDOX REACTIONS IN OCHEM

1. OXIDATION

- An **oxidation r**eaction is a chemical change in which the number of C-Y bonds (Y = O, halogen) **has increased** or a reaction in which the number of C-H bonds **has decreased**.

Ex:

416

2. REDUCTION

- When the number of C-Y bonds (Y = O, halogen) has **decreased** in a chemical process, a **reduction** reaction has occurred. Reduction also occurs when the number of C-H bonds has **increased**.

Ex:

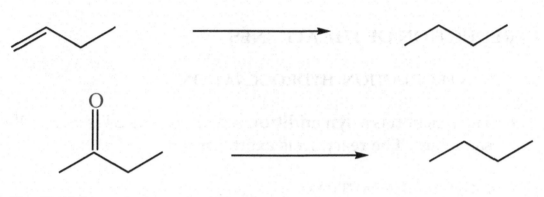

- Do Problem _____, page _____.

C. COMMON REDUCING AGENTS IN OCHEM

1. TYPES OF REDUCING AGENTS: 3

a. H_2/metal

b. $H_2 = 2H^+ + 2e^-$

Ex: Na/NH_3 or Li/NH_3

$$2Na \longrightarrow 2Na^+ + 2e^-$$

$$2NH_3 \longrightarrow 2NH_2^- + 2H^+$$

The overall reaction is:

$$2Na + 2NH_3 \longrightarrow 2Na^+ + 2NH_2^- + 2H^+ + 2e^-$$

c. Addition of a hydride and a proton

- There are **2 reducing** agents:

    o -LiAlH₄: **Lithium aluminum hydride**

    o -NaBH₄: **sodium borohydride**

D. REDUCTION OF THE ALKENES

1. INTRODUCTION: HYDROGENATION

- The reaction is a **syn addition reaction.** The 2 Hs add to the same face. **The reaction is exothermic.**

2. GENERAL REACTION

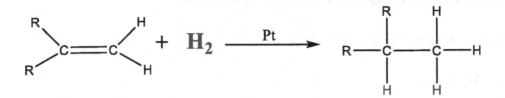

Ex:

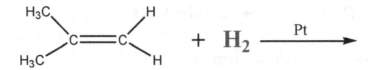

- Mechanism: See page _____.

3. HYDROGENATION AND STABILITY

- The more stable the alkene, the lower the heat of hydrogenation. **Indeed, the heats of hydrogenation can be used to assess alkene stability as follows:**

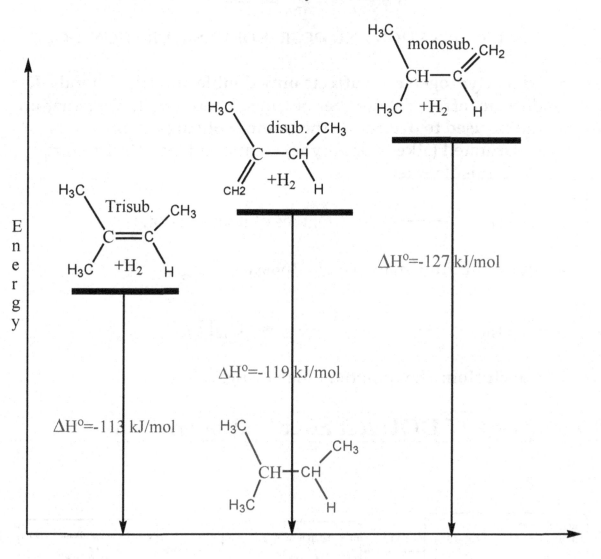

- The rate of the reaction increases as the number of R groups decreases. Indeed, the rate of the hydrogenation reaction follows the following order: monosubstituted alkenes > disubstituted alkenes > trisubs > tetrasub (very hindered). **As a result, the increasing order of stability of the alkenes is: monosubstituted alkenes < disubstituted alkenes < trisubs < tetrasub.**

- Read pages _____ - _____.

- See Fig. _____, page _____.

- Do problems on pages _____ - _____.

4. HYDROGENATION AND DEGREE OF UNSATURATION (DOU)

- Note: Hydrogenation affects only double and triple bonds. It does not affect the number of rings. Therefore, hydrogenation can be used to determine the number of rings in an unsaturated (alkene or alkyne) compound. See Unit 13 for DOU calculations.

Ex:

$$C_{10}H_{16} \longrightarrow 3 \text{ degrees of unsaturation}$$

- **Question: How many rings? 3possible rings.**

$$C_{10}H_{16} \xrightarrow{\text{hydrogenation}} C_{10}H_{20}$$

- **Conclusion**: The compound has 1 ring.

---

rings = # of DOU left after hydrogenation

or

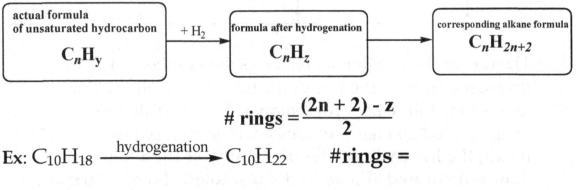

| actual formula of unsaturated hydrocarbon C_nH_y | $+ H_2$ | formula after hydrogenation C_nH_z | | corresponding alkane formula C_nH_{2n+2} |

$$\text{\# rings} = \frac{(2n + 2) - z}{2}$$

Ex: $C_{10}H_{18} \xrightarrow{\text{hydrogenation}} C_{10}H_{22}$ **#rings =**

- Read page _____ - _____ and associated problem.

5. HYDROGENATION OF OILS: HARDENING

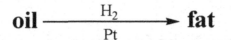

$$\text{oil} \xrightarrow[\text{Pt}]{\text{H}_2} \textbf{fat}$$

- Oils = triacylglycerols with a significant number of unsaturation
- Fats = triacylglycerols with few degrees of unsaturation. They have higher MP due to **increased surface area**. One can go from an oil to a fat through hydrogenation.

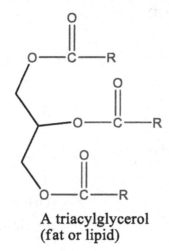

A triacylglycerol
(fat or lipid)

If R, R', and R" have many unsaturations = oil

If R, R', and R" have only few unsaturations = fat

- **Read pages _____ – _____ and associated problem.**

6. OXYMERCURATION-REDUCTION OF THE ALKENES

a. Introduction

- The reaction proceeds through the formation of a cyclic mercurinium ion in the first step with mercuric acetate, $Hg(OAc)_2$.
- The water attacks at the most substituted carbon of the mercurinium intermediate; Markovnikov's rule is followed.
- $NaBH_4$ is used as a reducing agent in the final step.
- An alcohol is produced. OAc = Acetate.

b. The General Reaction

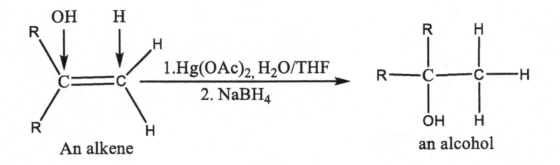

Ex:

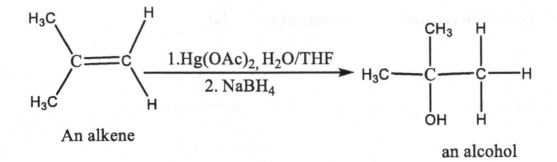

c. Mechanism of the Reaction

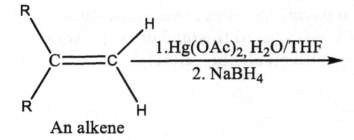

An alkene

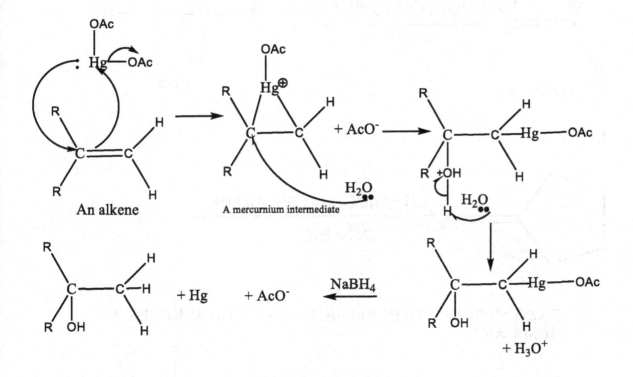

An alkene   A mercurnium intermediate

423

d. Alkoxymercuration-Reduction of the Alkenes

- The reaction is similar to oxymercuration, except mercuric trifluoroacetate ($Hg(O_2CCF_3)_2$ is used (instead of $Hg(OAc)_2$) and the water is replaced with an alcohol, ROH.
- An ether is produced.

- The general reaction is:

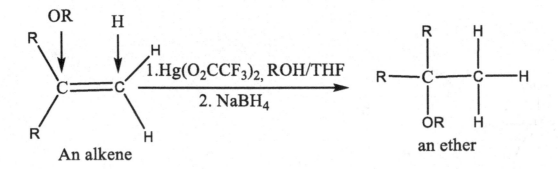

Ex:

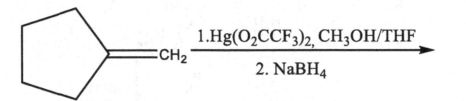

7. OXYMERCURATION-REDUCTION OF THE ALKENES: A SUMMARY

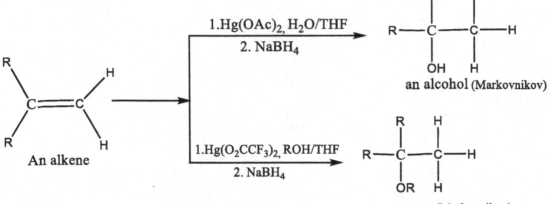

E. REDUCTION OF THE ALKYNES: HYDROGENATION

1. INTRODUCTION

- There are **3 ways** to hydrogenate an alkyne:

    o **Complete hydrogenation with H₂/Pd or H₂/Pt.**

    o **Partial hydrogenation with Lindlar catalyst.**

    o **Partial reduction using Li/NH₃ or Na/NH₃.**

2. COMPLETE REDUCTION

- Complete hydrogenation requires **2 moles of H₂ (or excess H₂)**. The reaction is a **syn addition reaction**.

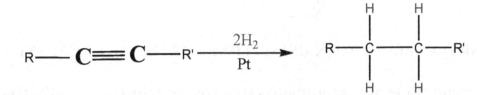

Ex:

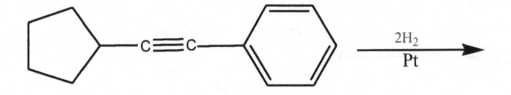

3. PARTIAL REDUCTION WITH LINDLAR CATALYST

- The reaction is **stereoselective (syn addition). Only the cis isomer is formed.**

- Note: Lindlar catalyst: Pd on $CaCO_3$ in $Pb(CH_3COO)_2$ and quinoline. Lindlar catalyst is also called partially deactivated palladium.

- The general reaction is:

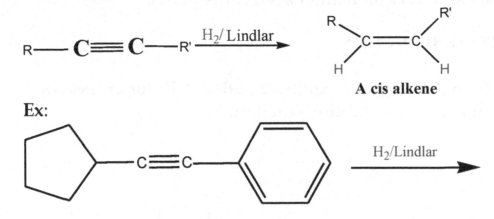

A cis alkene

Ex:

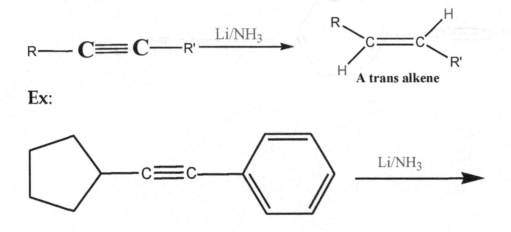

4. PARTIAL REDUCTION USING Li/NH₃ OR Na/NH₃ CATALYST

- The reaction is also **stereoselective (trans addition). Only the trans isomer is observed.**

- The general reaction is:

$$R \longrightarrow C \equiv C \longrightarrow R' \xrightarrow{Li/NH_3}$$

A trans alkene

Ex:

- See mechanism on page _____ and do associated problem.

- See summary on page _____ Fig. _____ and do associated problems.

5. REDUCTION OF THE ALKYNES: A SUMMARY

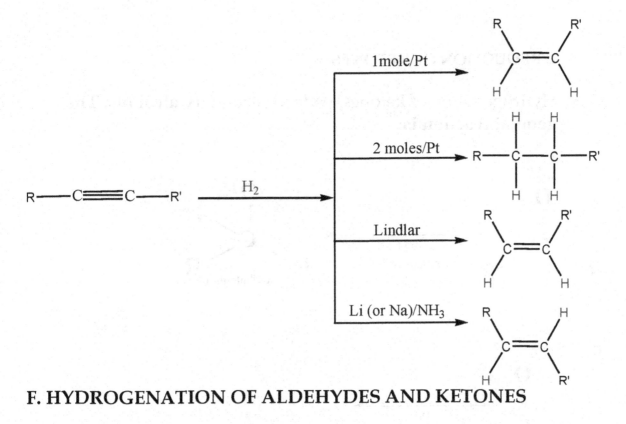

F. HYDROGENATION OF ALDEHYDES AND KETONES

1. REDUCTION OF ALDEHYDES

- Reduction of aldehydes with hydrogen leads to **primary alcohols as follows:**

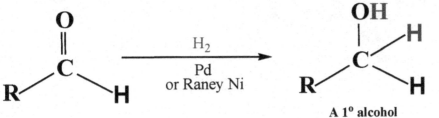

A 1° alcohol

- **Note: Raney Ni is finely dispersed Ni with adsorbed H_2.**

Ex:

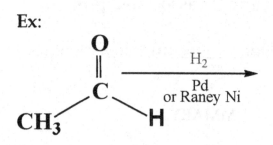

2. REDUCTION OF KETONES

- Hydrogenation of ketones leads to **secondary alcohols. The general reaction is:**

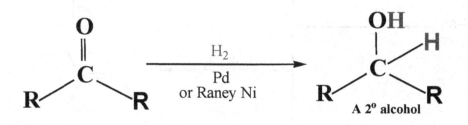

Ex:

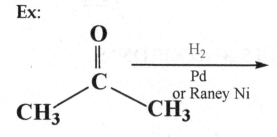

3. REDUCTION OF ALDEHYDES, KETONES, AND ACID CHLORIDES: A SUMMARY

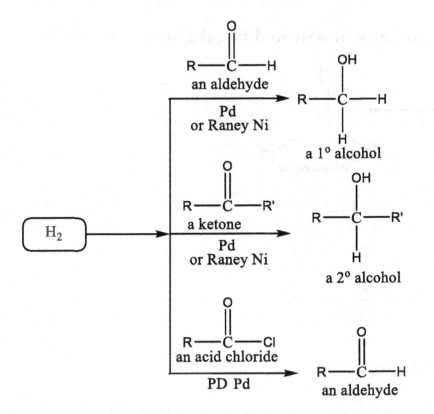

G. REDUCTION OF ALKYL HALIDES AND ACID CHLORIDES

1. REDUCTION OF ALKYL HALIDES

- The catalyst is $LiAlH_4$. The rate of the reaction decreases from 1º to 3º alkyl halides.

$$R\text{-}X \xrightarrow[\text{2. }H_2O]{\text{1. }LiAlH_4} \textbf{R-H}$$

An alkane

Ex:

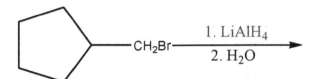

2. REDUCTION OF ACID CHLORIDES: THE ROSENMUND REDUCTION

- The catalyst is **partially deactivated Pd** (aka Lindlard Catalyst).

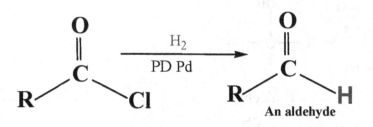

Ex:

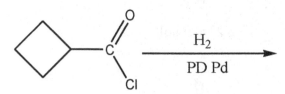

H. REDUCTION OF EPOXIDES

- This is an SN2 reaction in which **H⁻ (from LiAlH₄) is the nucleophile.**

- **Note: For unsymmetrical epoxides, the nucleophilic attack occurs at the least substituted carbon.**

- The general reaction is:

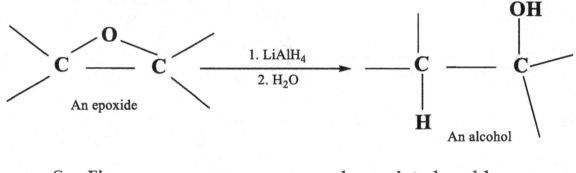

- See Fig. _____, page _____ and associated problems.

Ex:

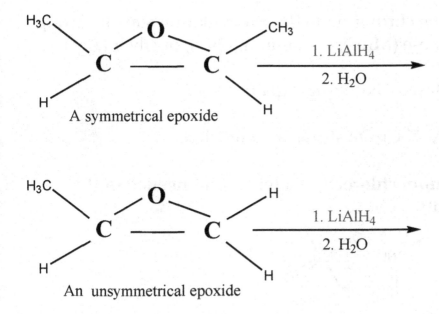

A symmetrical epoxide

An unsymmetrical epoxide

I. OXIDIZING AGENTS

1. INTRODUCTION

- There are **two types** of oxidizing agents:
    - -The first group consists of reagents having O-O bonds (Ex: peroxides, O_2, O_3).
    - -The other group of reagents contains metal-O bonds (Ex: CrO_4^{2-}).

2. O-O bond oxidizing agents

- Some are **O_2, O_3, H_2O_2, tert-butyl hydroperoxides ((CH_3)$_3$COOH), peroxyacids (RCO_3H).**

- See Fig. on page _____.

3. METAL-O BOND OXIDIZING AGENTS

a. Introduction

- The metal can be **chromium** in the **+ 6 oxidation state** in **strong acid** or manganese (Mn^{+7}), or osmium (Os^{+8}), or silver (Ag^+) .

b. Cr-Based Oxidizing Agents

- CrO_3, $Na_2Cr_2O_7$, $K_2Cr_2O_7$: **strong acid needed.**

- **PCC** = **pyridinium chlorochromate**: no acid needed (mild oxidizing agent).

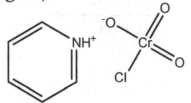

pyridinium chlorochromate

c. Mn-Based Oxidizing Agent = $KMnO_4$

d. Osmium-Based Oxidizing Agent = OsO_4

e. Ag-Based Oxidizing Agent = Ag_2O

J. OXIDATION OF ALKENES

1. EPOXIDATION

a. General Reaction: See Fig. _____, page _____.

- The reaction is carried out with a **peroxyacid** (RCO_3H).

- **The general reaction is:**

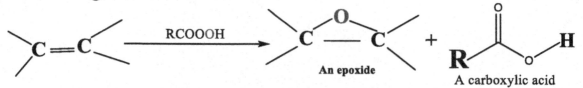

432

Ex:

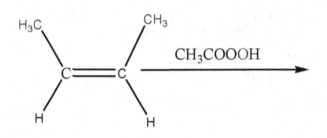

- See Fig. on page _____ for the structures of peroxyacids (PAA, mCPBA, MMPP).

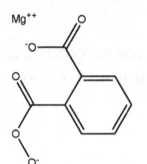

peroxyacetic acid **(PAA)**

magnesium monoperoxyphtalate **(MMPP)**

meta-chloroperoxybenzoic acid **(mCPBA)**

- See mechanism of the reaction on page _____.

- See examples on page _____.

- Do Problem _____, page _____.

b. Stereochemistry of Alkene Epoxidation

i. Epoxidation: a Syn Addition

- The reaction occurs through a **syn addition**. The configuration of the alkene is retained.

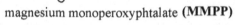

- o **cis-alkene ⟶ cis-epoxide**
- o **trans-alkene ⟶ trans-epoxide**

Ex:

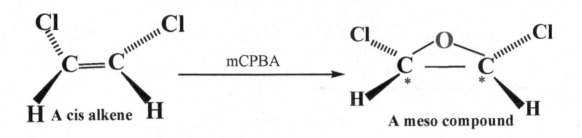

ii. Chirality of Epoxidation of the Alkene

- The reaction is **stereospecific.**

  - ○ **cis-alkene ⟶ an achiral meso compound**
  - ○ **trans-alkene ⟶ 2 enantiomers: A racemate**

Ex:

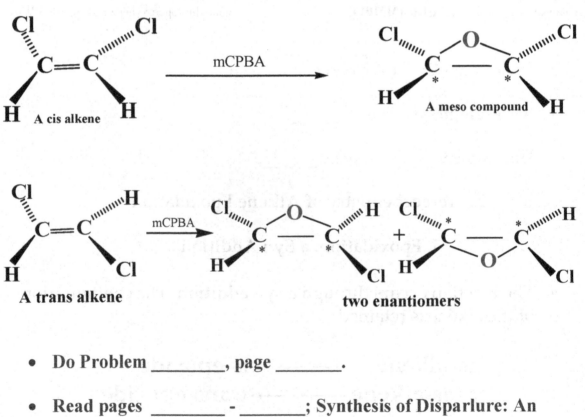

- **Do Problem _____, page _____.**

- **Read pages _____-_____; Synthesis of Disparlure: An Application of Epoxidation on page _____.**

2. OXIDATION OF THE ALKENES: DIHYDROXYLATION

a. Introduction

- **The general reaction is:**

Ex:

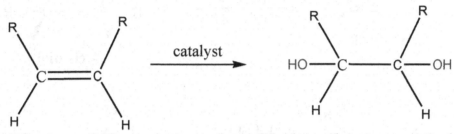

- **Note: The two OH groups can be added on the same side (syn dihydroxylation) or on opposite sides (anti dihydroxylation).**

b. Anti Dihydroxylation of the Alkenes

- The reaction proceeds in **two steps** leading to a **trans-1,2-diol and its enantiomer. A racemic mixture is obtained.** A **peroxyacid catalyst** is used.

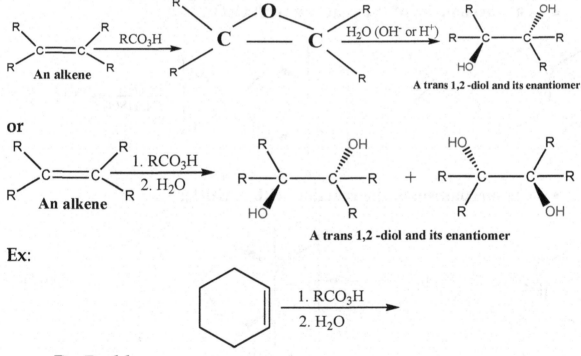

or

Ex:

- **Do Problem** _____ **on page** _____ .

c. Syn Dihydroxylation of the Alkenes

- The catalysts: **alkaline cold dilute $KMnO_4$ or $O.sO_4/NaHSO_3$, H_2O (H_3O^+). The general reaction is:**

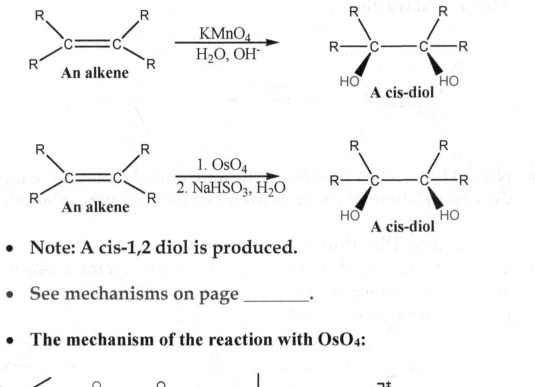

or

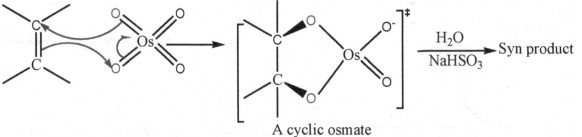

- **Note: A cis-1,2 diol is produced.**

- **See mechanisms on page _____.**

- **The mechanism of the reaction with OsO_4:**

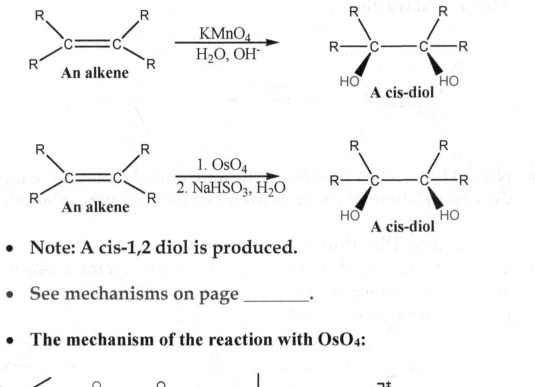

A cyclic osmate

- **The mechanism of the reaction with $KMnO_4$:**

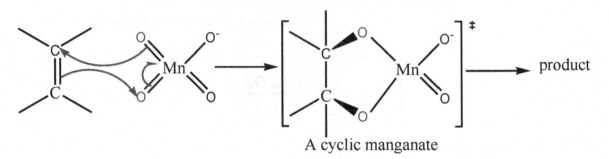

A cyclic manganate

Ex:

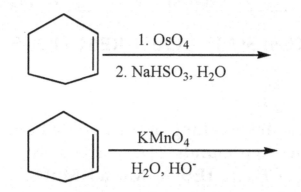

or

1. OsO₄

2. NaHSO₃, H₂O

KMnO₄

H₂O, HO⁻

d. Syn Dihydroxylation of the Alkenes: Using the OsO₄/NMO catalyst

- **OsO₄** is **a more selective oxidizing agent** than $KMnO_4$, but it is toxic and expensive. So a small amount of osmium tetroxide is combined with NMO (**N-methylmorpholine N-oxide**), a good oxidizing agent. In this "tandem" catalysis, Os(VIII) is reduced by the π electrons of the alkene double bond to Os(VI). Then, NMO oxidizes Os(VI) back to Os(VIII) which can be used again. The structure of **NMO** is:

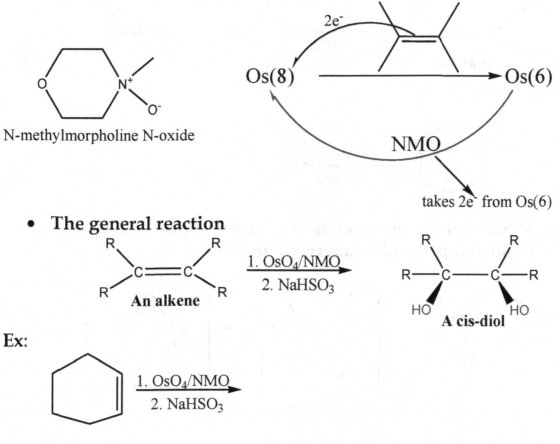

N-methylmorpholine N-oxide

$2e^-$

Os(8) ⟶ Os(6)

NMO

takes 2e⁻ from Os(6)

- **The general reaction**

R₂C=CR₂

An alkene

1. OsO₄/NMO

2. NaHSO₃

A cis-diol

Ex:

1. OsO₄/NMO

2. NaHSO₃

- **Read pages** _____ – _____. **Do problem** _____ **on page** _____.

K. OXIDATION OF ALKENES: CLEAVAGE REACTIONS

1. INTRODUCTION

- **A cleavage** reaction is the **breaking of the double bond.** The products are **carbonyl compounds.** The **catalysts** are $O_3/Zn/H_3O^+$, O_3/CH_3SCH_3 **(or H₂S) or warm KMnO₄.**

2. CLEAVAGE WITH OZONE: OZONOLYSIS

- **The general reaction is:**

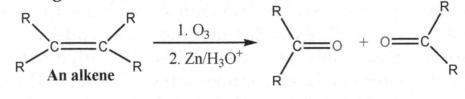

or

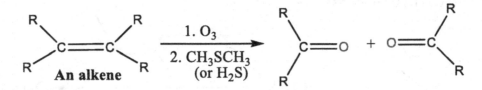

Ex:

- **Mechanism of the reaction: The reaction proceeds through molozonide/ozonide intermediates. See page** _____.

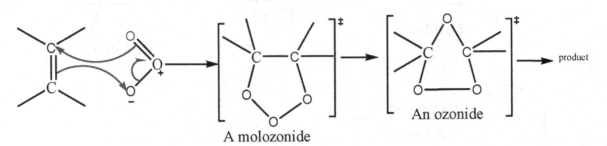

A molozonide   An ozonide

438

Ex:

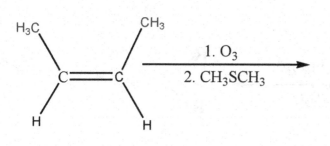

$$\xrightarrow[\text{2. CH}_3\text{SCH}_3]{\text{1. O}_3}$$

- **Note: The carbonyl products are <u>not</u> further oxidized to carboxylic acids.**

- **Do example and problem on pages _____ - _____.**

3. PRODUCTS IN THE CLEAVAGE OF ALKENES WITH ACIDIC OR NEUTRAL KMnO₄

- The product of the reaction depends on the alkene:
  - If there are **Hs** on the carbons of the double bond, **2 carboxylic acids** are formed.
  - If there are **2 Hs** on **a** carbon of the double bond, **CO₂** is formed from that carbon.
  - If there is **no H** present on the double bond, **two ketones** are formed.

- **The general reaction is:**

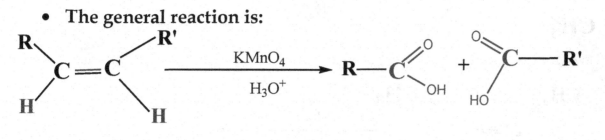

or

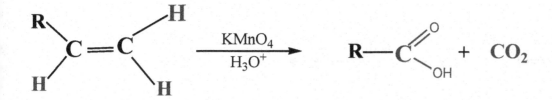

or

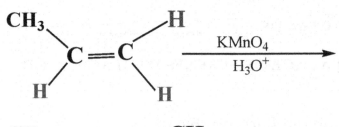

Ex:

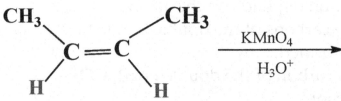

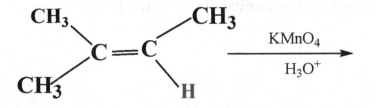

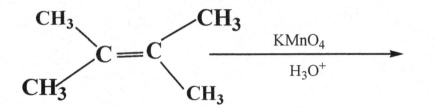

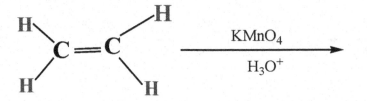

440

4. OXIDATION OF THE ALKENES: A SUMMARY

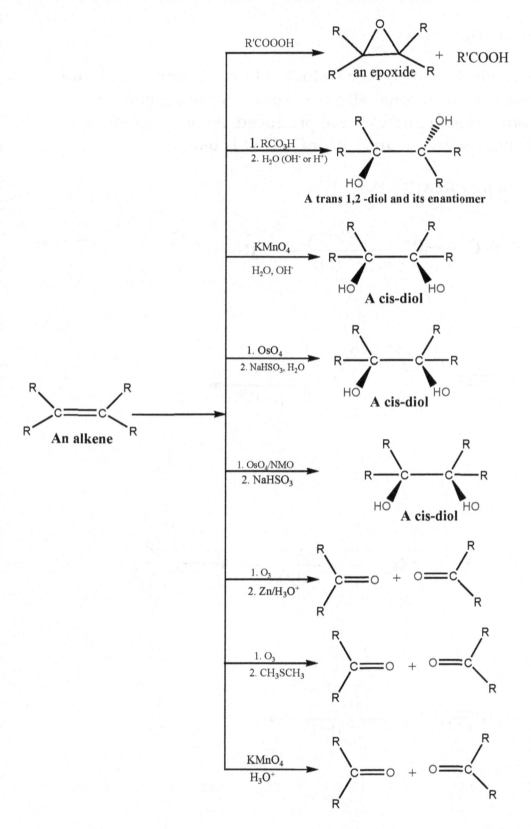

L. OXIDATION OF ALKYNES: CLEAVAGE REACTIONS

1. INTRODUCTION

- The catalysts: O_3/H_2O or $KMnO_4/H_3O^+$. Carbonic acids are produced for internal alkynes. For terminal alkynes, a carboxylic acid and CO_2 are produced. An α-diketone is produced when neutral $KMnO_4/H_2O$ is used.

2. GENERAL REACTIONS

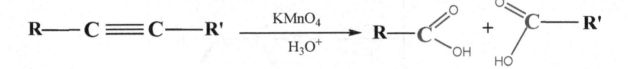

Ex:

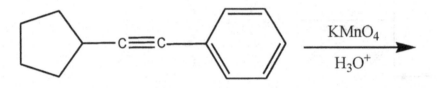

or

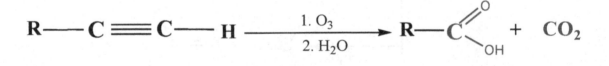

Ex:

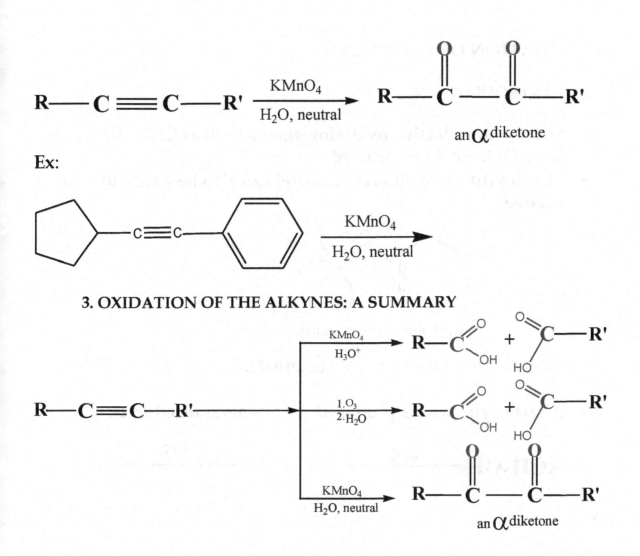

R—C≡C—R' $\xrightarrow[\text{H}_2\text{O, neutral}]{\text{KMnO}_4}$ R—$\overset{\overset{\displaystyle O}{\|}}{C}$—$\overset{\overset{\displaystyle O}{\|}}{C}$—R'

an α diketone

Ex:

3. OXIDATION OF THE ALKYNES: A SUMMARY

- **Note: Oxidation can be used to locate a multiple bond in an unknown unsaturated compound.**

Ex: What is the **structural formula** of the reactant in the following reaction?

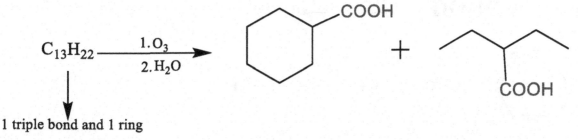

$C_{13}H_{22}$ $\xrightarrow[\text{2. H}_2\text{O}]{\text{1. O}_3}$

1 triple bond and 1 ring

- **Read page _____ – _____ and do appropriate problems.**

- **Do example , page _____.**

M. OXIDATION OF ALCOHOLS

1. INTRODUCTION

- Strong, nonselective oxidizing agents such as CrO_3, $Na_2Cr_2O_7$, $K_2Cr_2O_7$ in acid can be used.
- PCC (pyridinium chlorochromate) can also be used: no acid needed)

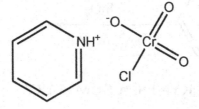

pyridinium chlorochromate

2. OXIDATION OF PRIMARY ALCOHOLS

- A carboxylic acid is produced. The general reaction is:

$$RCH_2OH \xrightarrow{\ [O]\ } \qquad\qquad \xrightarrow{\ [O]\ }$$

Ex:

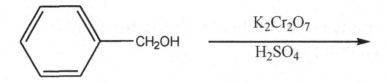

$$RCH_2OH \xrightarrow{\ [PCC]\ }$$

Ex:

- See mechanism of reaction on page _____.

- Read about the Breathalyzer. See Fig. _____, page _____.

444

3. OXIDATION OF 2⁰ ALCOHOLS

- **A ketone is produced. The general reaction is:**

$$R_2CHOH \xrightarrow{[O]}$$

Ex:

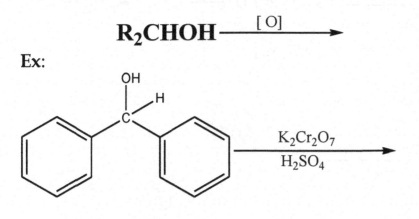

4. OXIDATION OF 3⁰ ALCOHOLS
- **No reaction occurs.**

$$R_3COH \xrightarrow{[O]} \textbf{No Reaction}$$

Ex:

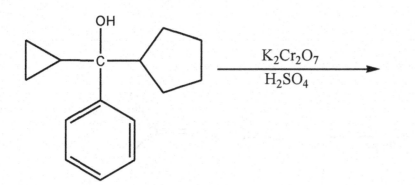

- Read about the oxidation of ethanol, alcoholism and Antabuse, methanol poisoning and treatment, on pages _____ - _____.

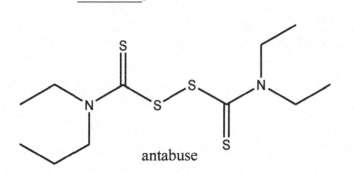

antabuse

5. OXIDATION THE ALCOHOLS: A SUMMARY

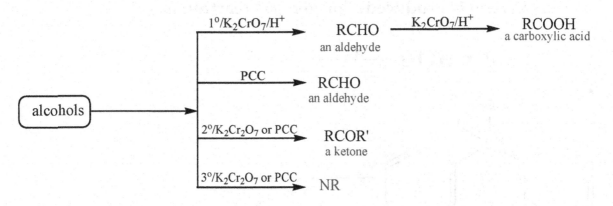

N. OXIDATION OF 1,2 – DIOLS

1. INTRODUCTION

- 1,2 - diols can be cleaved **using aqueous periodic acid in THF. Two carbonyl compounds are produced. The general reaction is:**

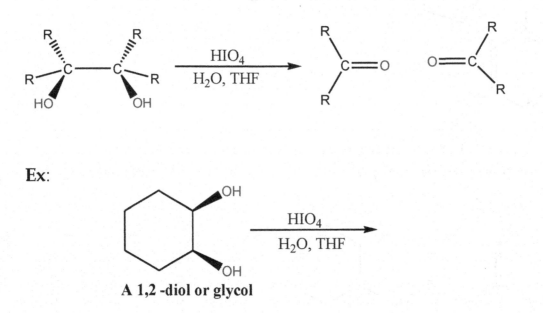

Ex:

- **Mechanism of the reaction:**

- **The reaction proceeds through a cyclic periodate intermediate.**

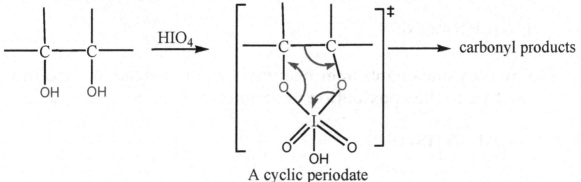

A cyclic periodate

Ex:

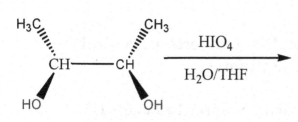

2. INDIRECT CLEAVAGE: COUPLING OsO₄ WITH HIO₄

- **The general reaction is:**

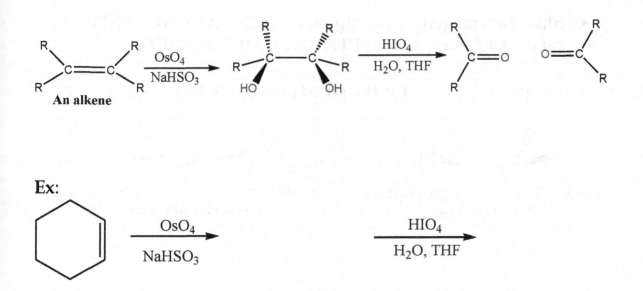

An alkene

Ex:

O. SHARPLESS EPOXIDATION

- Named after **K. Barry Sharpless of Scripps Research Institute,** Nobel Prize in Chemistry in 2001.

  ## 1. THE PROBLEM

- Can only one enantiomer be mostly made instead of a racemic mixture in the epoxidation of alkenes? See page _____.

  ## 2. SOME DEFINITIONS

- Enantioselective reaction = a reaction that gives one predominant enantiomer.

- Asymmetric reaction = reaction that converts an achiral reactant to one predominant enantiomer.

  ## 4. CONVERTING ALLYLIC ALCOHOLS (special alkenes) TO EPOXIDES: AN OXIDATION REACTION

- See page _____.

- Read pages _____ – _____.

- Sharpless reagent = 3 components: $(CH_3)_3COOH$ + Ti(IV) catalyst + (+ or -) Diethyl Tartrate (+DET or –DET).

- See page _____ for the structures of DET.

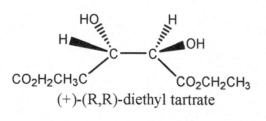
(+)-(R,R)-diethyl tartrate

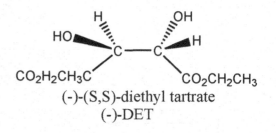
(-)-(S,S)-diethyl tartrate
(-)-DET

- With (-)-DET, the oxygen atom is added from above the plane. The top epoxide is the major product.

- With (+)-DET, the oxygen atom is added from below the plane. The bottom epoxide is the major product.
- The general reaction is:

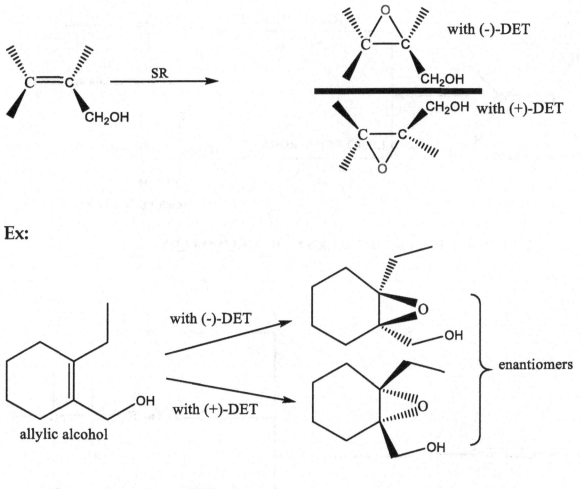

Ex:

- See examples on pages _____ - _____.

O. SOME SELECTED REACTIONS

1. OXYMERCURATION-REDUCTION OF THE ALKENES: A SUMMARY

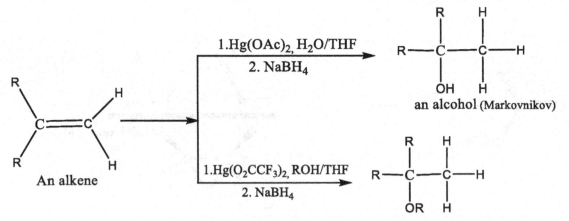

2. REDUCTION OF THE ALKYNES: A SUMMARY

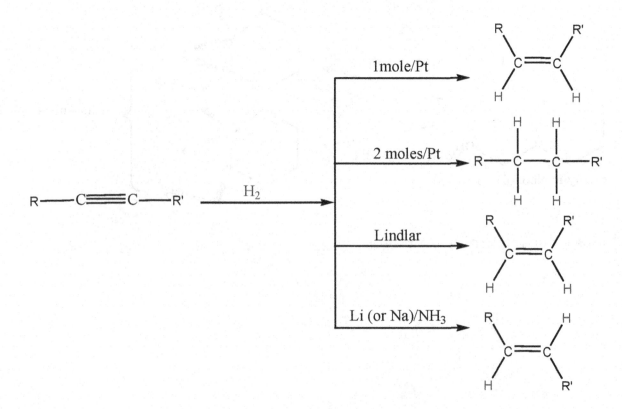

3. REDUCTION OF ALDEHYDES, KETONES, AND ACID CHLORIDES: A SUMMARY

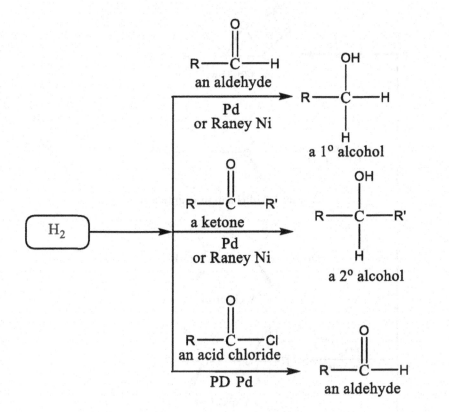

4. OXIDATION OF THE ALKYNES: A SUMMARY

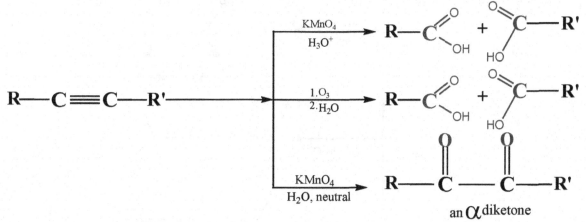

5. OXIDATION OF THE ALKENES: A SUMMARY

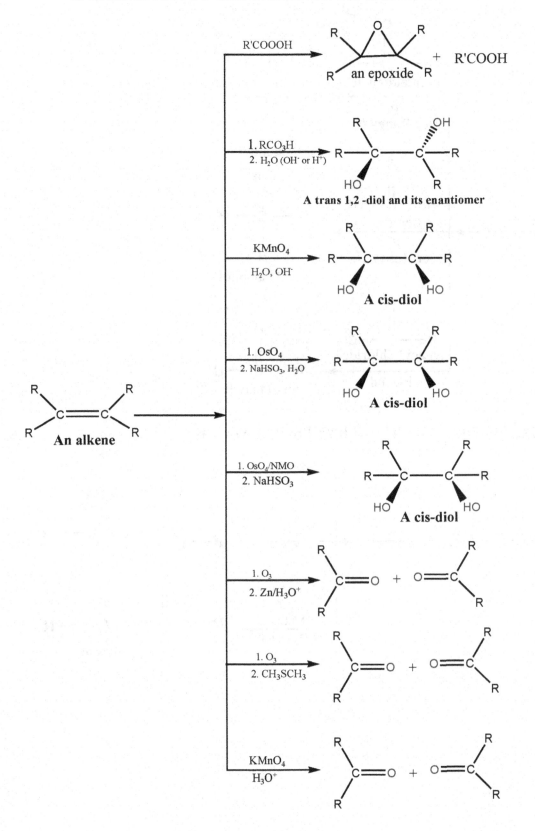

6. OXIDATION OF THE ALCOHOLS: A SUMMARY

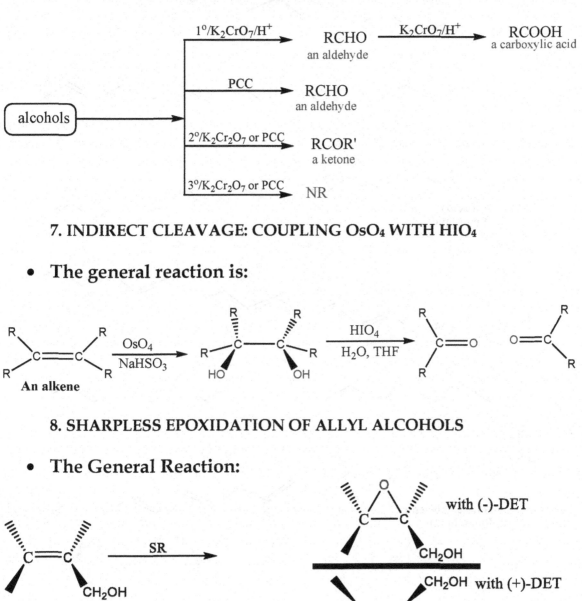

7. INDIRECT CLEAVAGE: COUPLING OsO₄ WITH HIO₄

- **The general reaction is:**

8. SHARPLESS EPOXIDATION OF ALLYL ALCOHOLS

- **The General Reaction:**

- **See Key Concepts on pages _____ – _____.**

453

• <u>Oils and Fats</u>

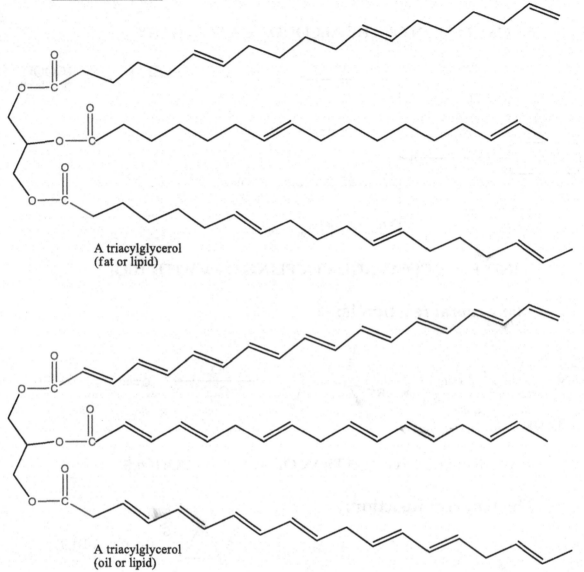

A triacylglycerol
(fat or lipid)

A triacylglycerol
(oil or lipid)